Geometry Station Activities
for Common Core Standards

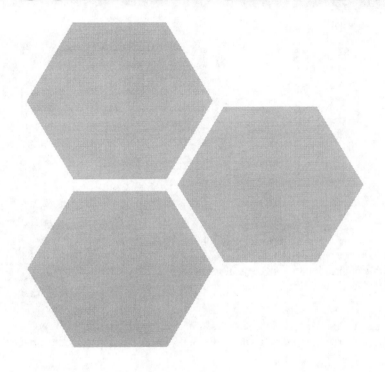

The classroom teacher may reproduce materials in this book for classroom use only.
The reproduction of any part for an entire school or school system is strictly prohibited.
No part of this publication may be transmitted, stored, or recorded in any form
without written permission from the publisher.

© Common Core State Standards for Mathematics. Copyright 2010. National Governor's
Association Center for Best Practices and Council of Chief State School Officers. All rights reserved.

1 2 3 4 5 6 7 8 9 10
ISBN 978-0-8251-6788-1
Copyright © 2011
J. Weston Walch, Publisher
Portland, ME 04103
www.walch.com
Printed in the United States of America

Table of Contents

Standards Correlations v
Introduction vii
Materials List x

Congruence 1
   Set 1: Parallel Lines and Transversals 1
   Set 2: Classifying Triangles and Angle Theorems 13
   Set 3: Corresponding Parts, Transformations, and Proof 28
   Set 4: Special Congruent Triangles 40
   Set 5: Bisectors, Medians, and Altitudes 50
   Set 6: Triangle Inequalities 65
   Set 7: Translations, Dilations, Tessellations, and Symmetry 77
   Set 8: Coordinate Proof 90
   Set 9: Rhombi, Squares, Kites, and Trapezoids 101

Similarity, Right Triangles, and Trigonometry 112
   Set 1: Similarity and Scale Factor 112
   Set 2: Ratio Segments 126
   Set 3: Sine, Cosine, and Tangent Ratios, and
         Angles of Elevation and Depression 139

Circles 151
   Set 1: Circumference, Angles, Arcs, Chords, and Inscribed Angles 151
   Set 2: Special Segments, Angle Measurements, and Equations of Circles 165
   Set 3: Circumcenter, Incenter, Orthocenter, and Centroid 177

Expressing Geometric Properties with Equations 192
   Set 1: Parallel Lines, Slopes, and Equations 192
   Set 2: Perpendicular Lines 205
   Set 3: Coordinate Proof with Quadrilaterals 219

Standards Correlations

The standards correlations below support the implementation of the Common Core Standards. *Geometry Station Activities for Common Core State Standards* includes station activity sets for the high school Common Core geometry domains of Congruence; Similarity, Right Triangles, and Trigonometry; Circles; and Expressing Geometric Properties with Equations. This table provides a listing of the available station activities organized by Common Core standard.

The left column lists the standard codes. The first letter of the code represents the Common Core grade level standard for high school. This letter is followed by a dash and the initials of the domain name, which is then followed by the standard number. The middle column lists the title of the station activity set that corresponds to the standard, and the right column lists the page number where the station activity set can be found.

The table indicates the standards that are heavily addressed in the station sets. If there are other standards that are addressed within the set, they can be found on the first page of each set.

Standard	Set title	Page number
G-C0.2.	Corresponding Parts, Transformations, and Proof	28
G-C0.3.	Translations, Dilations, Tessellations, and Symmetry	77
G-C0.3.	Coordinate Proof	90
G-CO.4.	Translations, Dilations, Tessellations, and Symmetry	77
G-CO.5.	Corresponding Parts, Transformations, and Proof	28
G-CO.5.	Translations, Dilations, Tessellations, and Symmetry	77
G-CO.5.	Coordinate Proof	90
G-CO.6.	Corresponding Parts, Transformations, and Proof	28
G-CO.6.	Translations, Dilations, Tessellations, and Symmetry	77
G-CO.6.	Coordinate Proof	90
G-CO.7.	Corresponding Parts, Transformations, and Proof	28
G-CO.7.	Special Congruent Triangles	40
G-CO.8.	Corresponding Parts, Transformations, and Proof	28
G-CO.9.	Parallel Lines and Transversals	1
G-CO.10.	Classifying Triangles and Angle Theorems	13
G-CO.10.	Bisectors, Medians, and Altitudes	50
G-CO.10.	Triangle Inequalities	65
G-CO.10.	Ratio Segments	126
G-C0.11.	Rhombi, Squares, Kites, and Trapezoids	101

(*continued*)

Standards Correlations

Standard	Set title	Page number
G-C0.12.	Bisectors, Medians, and Altitudes	50
G-C0.12.	Rhombi, Squares, Kites, and Trapezoids	101
G-C0.12.	Ratio Segments	126
G-C0.12.	Circumcenter, Incenter, Orthocenter, and Centroid	177
G-CO.12.	Perpendicular Lines	205
G-CO.13.	Circumcenter, Incenter, Orthocenter, and Centroid	177
G-SRT.1.	Translations, Dilations, Tessellations, and Symmetry	77
G-SRT.1.	Coordinate Proof	90
G-SRT.2.	Translations, Dilations, Tessellations, and Symmetry	77
G-SRT.2.	Similarity and Scale Factor	112
G-SRT.4.	Ratio Segments	126
G-SRT.5.	Similarity and Scale Factor	112
G-SRT.5.	Ratio Segments	126
G-SRT.6.	Sine, Cosine, and Tangent Ratios, and Angles of Elevation and Depression	139
G-SRT.7.	Sine, Cosine, and Tangent Ratios, and Angles of Elevation and Depression	139
G-SRT.8.	Special Congruent Triangles	40
G-SRT.8.	Sine, Cosine, and Tangent Ratios, and Angles of Elevation and Depression	139
G-C.2.	Circumference, Angles, Arcs, Chords, and Inscribed Angles	151
G-C.2.	Special Segments, Angle Measurements, and Equations of Circles	165
G-C.3.	Circumcenter, Incenter, Orthocenter, and Centroid	177
G-C.4.	Special Segments, Angle Measurements, and Equations of Circles	165
G-C.5.	Circumference, Angles, Arcs, Chords, and Inscribed Angles	151
G-C.5.	Special Segments, Angle Measurements, and Equations of Circles	165
G-GPE.1.	Special Segments, Angle Measurements, and Equations of Circles	165
G-GPE.4.	Corresponding Parts, Transformations, and Proof	28
G-GPE.4.	Special Congruent Triangles	40
G-GPE.4.	Similarity and Scale Factor	112
G-GPE.4.	Coordinate Proof with Quadrilaterals	219
G-GPE.5.	Parallel Lines, Slopes, and Equations	192
G-GPE.5.	Perpendicular Lines	205
G-GPE.7.	Similarity and Scale Factor	112

Introduction

Geometry Station Activities for Common Core State Standards includes a collection of station-based activities to provide students with opportunities to practice and apply the mathematical skills and concepts they are learning. It contains several sets of activities for the following Common Core domains for high school geometry: Congruence; Similarity, Right Triangles and Trigonometry; Circles; and Expressing Geometric Properties with Equations. You may use these activities as a complement to your regular lessons or in place of your regular lessons, if formative assessment suggests students have the basic concepts but need practice. The debriefing discussions after each set of activities provide an important opportunity to help students reflect on their experiences and synthesize their thinking. It also provides an additional opportunity for ongoing, informal assessment to inform instructional planning.

Implementation Guide

The following guidelines will help you prepare for and use the activity sets in this book.

Setting Up the Stations

Each activity set consists of four stations. Set up each station at a desk, or at several desks pushed together, with enough chairs for a small group of students. Place a card with the number of the station on the desk. Each station should also contain the materials specified in the teacher's notes, and a stack of student activity sheets (one copy per student). Place the required materials (as listed) at each station.

When a group of students arrives at a station, each student should take one of the activity sheets to record the group's work. Although students should work together to develop one set of answers for the entire group, each student should record the answers on his or her own activity sheet. This helps keep students engaged in the activity and gives each student a record of the activity for future reference.

Forming Groups of Students

All activity sets consist of four stations. You might divide the class into four groups by having students count off from 1 to 4. If you have a large class and want to have students working in small groups, you might set up two identical sets of stations, labeled A and B. In this way, the class can be divided into eight groups, with each group of students rotating through the "A" stations or "B" stations.

Introduction

Assigning Roles to Students

Students often work most productively in groups when each student has an assigned role. You may want to assign roles to students when they are assigned to groups and change the roles occasionally. Some possible roles are as follows:

- Reader—reads the steps of the activity aloud
- Facilitator—makes sure that each student in the group has a chance to speak and pose questions; also makes sure that each student agrees on each answer before it is written down
- Materials Manager—handles the materials at the station and makes sure the materials are put back in place at the end of the activity
- Timekeeper—tracks the group's progress to ensure that the activity is completed in the allotted time
- Spokesperson—speaks for the group during the debriefing session after the activities

Timing the Activities

The activities in this book are designed to take approximately 15 minutes per station. Therefore, you might plan on having groups change stations every 15 minutes, with a two-minute interval for moving from one station to the next. It is helpful to give students a "5-minute warning" before it is time to change stations.

Since the activity sets consist of four stations, the above timeframe means that it will take about an hour and 10 minutes for groups to work through all stations. If this is followed by a 20-minute class discussion as described below, an entire activity set can be completed in about 90 minutes.

Guidelines for Students

Before starting the first activity set, you may want to review the following "ground rules" with students. You might also post the rules in the classroom.

- All students in a group should agree on each answer before it is written down. If there is a disagreement within the group, discuss it with one another.
- You can ask your teacher a question only if everyone in the group has the same question.
- If you finish early, work together to write problems of your own that are similar to the ones on the student activity sheet.
- Leave the station exactly as you found it. All materials should be in the same place and in the same condition as when you arrived.

Introduction

Debriefing the Activities

After each group has rotated through every station, bring students together for a brief class discussion. At this time you might have the groups' spokespersons pose any questions they had about the activities. Before responding, ask if students in other groups encountered the same difficulty or if they have a response to the question. The class discussion is also a good time to reinforce the essential ideas of the activities. The questions that are provided in the teacher's notes for each activity set can serve as a guide to initiating this type of discussion.

You may want to collect the student activity sheets before beginning the class discussion. However, it can be beneficial to collect the sheets afterward so that students can refer to them during the discussion. This also gives students a chance to revisit and refine their work based on the debriefing session.

Guide to Common Core Standards Annotation

As you use this book, you will come across annotation symbols included with the Common Core standards for several station activity sets. The following descriptions of these annotation symbols are verbatim from the Common Core State Standards Initiative Web site, at www.corestandards.org.

Symbol: ★

Denotes: Modeling Standards

Modeling is best interpreted not as a collection of isolated topics but rather in relation to other standards. Making mathematical models is a Standard for Mathematical Practice, and specific modeling standards appear throughout the high school standards indicated by a star symbol (★).

From http://www.corestandards.org/the-standards/mathematics/high-school-modeling/introduction/

Symbol: (+)

Denotes: College and Career Readiness Standards

The evidence concerning college and career readiness shows clearly that the knowledge, skills, and practices important for readiness include a great deal of mathematics prior to the boundary defined by (+) symbols in these standards.

From http://www.corestandards.org/the-standards/mathematics/note-on-courses-and-transitions/courses-and-transitions/

Introduction

Materials List

Station Sets

- cardboard triangle created from a triangle with the vertices (4, 4), (10, 4), and (6, 12) in the coordinate plane
- colored markers (red, blue, green, and black specifically)
- cork board
- drinking straws
- dry spaghetti noodles (specifically, six that are 1 inch, 2 inches, 2.5 inches, 3.5 inches, 4 inches, and 5 inches in length, as well as others not cut to length)
- graphing calculators
- plastic coffee can lid
- push pins
- rubber bands
- small and large piece of poster board
- tape measure
- toothpicks
- tracing paper
- white computer paper

Class Sets

- calculators
- compasses
- protractors
- rulers
- scissors

Ongoing Use

- graph paper
- index cards (prepared according to specifications in teacher notes for many of the station activities)
- tape

Congruence

Set 1: Parallel Lines and Transversals

Instruction

Goal: To provide opportunities for students to develop concepts and skills related to identify and use the relationships between special pairs of angles formed by parallel lines and transversals

Common Core Standards

Congruence

Experiment with transformations in the plane.

G-CO.1. Know precise definitions of angle, circle, perpendicular line, parallel line, and line segment, based on the undefined notions of point, line, distance along a line, and distance around a circular arc.

Prove geometric theorems.

G-CO.9. Prove theorems about lines and angles.

Student Activities Overview and Answer Key

Station 1

Students will be given graph paper, a ruler, and a protractor. Students will use the graph paper and ruler to model parallel lines cut by a transversal. They will use the protractor to find vertical angles. Then they will use the graph paper to model lines, which are not parallel, that are cut by a transversal. They will use the protractor to find vertical angles. They will realize that vertical angles can be found by both parallel and non-parallel lines cut by a transversal.

Answers

1. Answers will vary.
2. $\angle A \cong \angle C$; $\angle B \cong \angle D$
3. vertical angles
4. $\angle E \cong \angle H$; $\angle F \cong \angle G$
5. Answers will vary.
6. Answers will vary.
7. $\angle E \cong \angle H$; $\angle F \cong \angle G$
8. vertical angles
9. $\angle A \cong \angle D$; $\angle B \cong \angle C$
10. Answers will vary.
11. Both parallel and non-parallel lines have vertical angles when cut by a transversal.

Congruence
Set 1: Parallel Lines and Transversals

Instruction

Station 2

Students will be given graph paper, a ruler, and a protractor. Students will use the graph paper and ruler to model parallel lines cut by a transversal. They will use the protractor to find supplementary angles. Then they will use the graph paper to model lines, which are not parallel, that are cut by a transversal. They will use the protractor to find supplementary angles. They will describe which types of angles are supplementary angles when two lines are cut by a transversal.

Answers

1. $\angle A$ and $\angle B$; $\angle C$ and $\angle D$; $\angle E$ and $\angle F$; $\angle G$ and $\angle H$; $\angle A$ and $\angle C$; $\angle B$ and $\angle D$; $\angle E$ and $\angle G$; $\angle F$ and $\angle H$

2. Answers will vary.

3. $\angle A$ and $\angle B$; $\angle C$ and $\angle D$; $\angle E$ and $\angle F$; $\angle G$ and $\angle H$; $\angle A$ and $\angle C$; $\angle B$ and $\angle D$; $\angle E$ and $\angle G$; $\angle F$ and $\angle H$

4. Answers will vary.

5. adjacent angles; interior angles on the same side of the transversal but only when the transversal intersects parallel lines.

Station 3

Students will be given spaghetti noodles, a protractor, graph paper, and a ruler. Students will use the graph paper and spaghetti noodles to model parallel lines cut by a transversal. They will use the protractor to measure the angles created by the transversal. Then they will explore and define the exterior, interior, alternate exterior, and alternate interior angles created by the transversal.

Answers

1. 8 angles
2. interior
3. exterior
4. 1, 2, 7, 8
5. 3, 4, 5, 6
6. Answers will vary.
7. $m\angle 1 = m\angle 7$; $m\angle 2 = m\angle 8$; $m\angle 3 = m\angle 5$; $m\angle 4 = m\angle 6$
8. 4 and 6; 3 and 5
9. 1 and 7; 2 and 8
10. equal; equal

Congruence
Set 1: Parallel Lines and Transversals

Instruction

Station 4

Students will be given spaghetti noodles, a protractor, and two parallel lines cut by a transversal. Students will use the spaghetti noodles to model the letter "F" to find corresponding angles. Then they will use the protractor to measure the angles and justify their answer.

Answers

1. 3 and 7
2. 1 and 5; 2 and 6; 3 and 7; 4 and 8
3. Answers will vary.
4. Answers will vary.

Materials List/Setup

Station 1   graph paper; ruler; protractor
Station 2   graph paper; ruler; protractor
Station 3   dry spaghetti noodles; protractor; graph paper; ruler
Station 4   dry spaghetti noodles; protractor

Congruence
Set 1: Parallel Lines and Transversals

Instruction

Discussion Guide

To support students in reflecting on the activities and to gather some formative information about student learning, use the following prompts to facilitate a class discussion to "debrief" the station activities.

Prompts/Questions

1. How many angles are created when parallel or non-parallel lines are cut by a transversal?
2. What are exterior angles?
3. What are interior angles?
4. What is a vertical angle?
5. What are alternate exterior angles?
6. What are alternate interior angles?
7. What two types of angles are supplementary when parallel lines are cut by a transversal?

Think, Pair, Share

Have students jot down their own responses to questions, then discuss with a partner (who was not in their station group), and then discuss as a whole class.

Suggested Appropriate Responses

1. 8 angles
2. Exterior angles lie on the outside of the lines cut by the transversal.
3. Interior angles lie in between the two lines cut by the transversal.
4. Vertical angles are two angles formed by two intersecting lines that lie on opposite sides of the point of intersection.
5. Alternate exterior angles are pairs of angles on opposite sides of the transversal that are outside of the parallel or non-parallel lines.
6. Alternate interior angles are pairs of angles on opposite sides of the transversal that are inside the parallel or non-parallel lines.
7. Interior angles on the same side of the transversal and adjacent angles are supplementary in this situation.

Congruence
Set 1: Parallel Lines and Transversals

Instruction

Possible Misunderstandings/Mistakes

- Not finding corresponding angles correctly and identifying that they have the same measure if the lines cut by the transversal are parallel
- Mixing up interior and exterior angles
- Not realizing vertical angles are always equal whether or not the lines cut by the transversal are parallel
- Not realizing that alternate interior or exterior angles must be on the opposite side of the transversals

NAME: _____

Congruence
Set 1: Parallel Lines and Transversals

Station 1

At this station, you will find graph paper, a ruler, and a protractor. As a group, construct two parallel lines that are cut by a transversal and label the angles as shown in the diagram below.

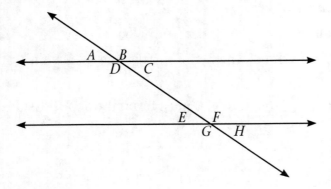

1. Use the protractor to measure the following angles:

   $m \angle A =$ _____ $m \angle B =$ _____ $m \angle C =$ _____ $m \angle D =$ _____

2. Which pairs of angles are equal in problem 1?

   _____

3. The pairs of angles found in problem 2 have a special name. What are these angles called? (*Hint:* Think about their location in relation to each other.)

   _____

4. What pairs of angles are congruent for angles *E*, *F*, *G*, and *H*?

   _____

5. What strategy did you use to find the angle pairs in problem 4?

continued

6

Geometry Station Activities for Common Core State Standards   © 2011 Walch Education

NAME: _____

Congruence
Set 1: Parallel Lines and Transversals

On your graph paper, construct a new graph of two lines that are NOT parallel. These lines are cut by a transversal. Label the angles as shown in the diagram below.

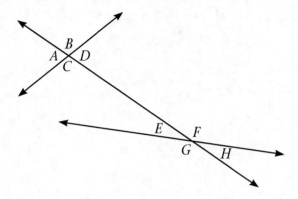

6. Use the protractor to measure the following angles:

   $m\angle E =$ _____ $m\angle F =$ _____ $m\angle G =$ _____ $m\angle H =$ _____

7. Which pairs of angles are equal?

8. The pairs of angles found in problem 7 have a special name. What are these angles called? (*Hint:* Think about their location in relation to each other.)

   _____

9. What pairs of angles are congruent for angles *A*, *B*, *C*, and *D*?

   _____

10. What strategy did you use to find the angle pairs in problem 9?

11. Based on your observations in problems 1–10, vertical angles can be found for what types of lines cut by a transversal?

NAME:

Congruence
Set 1: Parallel Lines and Transversals

Station 2

At this station, you will find graph paper, a ruler, and a protractor. As a group, construct two parallel lines that are cut by a transversal and label the angles as shown in the diagram below.

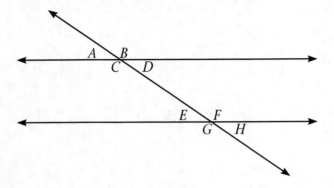

1. Which angles are supplementary angles?

2. Use your protractor to justify your answer to problem 1 by recording the measurements of each angle below.

   $m\angle A =$ _____ $m\angle C =$ _____ $m\angle E =$ _____ $m\angle G =$ _____

   $m\angle B =$ _____ $m\angle D =$ _____ $m\angle F =$ _____ $m\angle H =$ _____

On your graph paper, construct a new graph of two lines that are NOT parallel. These lines are cut by a transversal. Label the angles as shown in the diagram below.

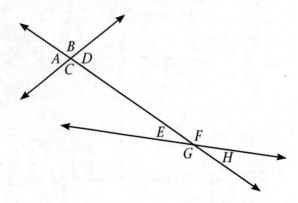

3. Which angles are supplementary angles?

continued

8

Geometry Station Activities for Common Core State Standards   © 2011 Walch Education

Congruence
Set 1: Parallel Lines and Transversals

4. Use your protractor to justify your answer to problem 3 by recording the measurements of each angle below.

   $m\angle A = $ _____ $m\angle C = $ _____ $m\angle E = $ _____ $m\angle G = $ _____

   $m\angle B = $ _____ $m\angle D = $ _____ $m\angle F = $ _____ $m\angle H = $ _____

5. Based on your answers and observations in problems 1–4, which of the following types of angles are supplementary?

   vertical angles   adjacent angles   alternate interior angles

   alternate exterior angles   interior angles on the same side of the transversal

Congruence

Set 1: Parallel Lines and Transversals

Station 3

At this station, you will find spaghetti noodles, a protractor, graph paper, and a ruler. Follow the directions below, and then answer the questions.

- On the graph paper, construct two parallel horizontal lines.
- Construct a diagonal line that passes through both parallel lines. This line is called a transversal.

1. How many angles does this transversal create with the two parallel lines?

   _____

On your graph paper, label the angles to model the parallel lines and transversal shown below.

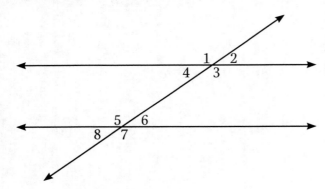

Place a spaghetti noodle on each parallel line.

2. Are the angles between the spaghetti noodles called exterior or interior angles?

   _____

3. Are the angles above the top parallel spaghetti noodle and below the bottom spaghetti noodle called exterior or interior angles?

   _____

continued

Congruence
Set 1: Parallel Lines and Transversals

4. Which angles are exterior angles? _____

5. Which angles are interior angles? _____

6. Use your protractor to measure each angle. Write the angle measurements below.

    $m\angle 1 =$ _____ $m\angle 3 =$ _____ $m\angle 5 =$ _____ $m\angle 7 =$ _____

    $m\angle 2 =$ _____ $m\angle 4 =$ _____ $m\angle 6 =$ _____ $m\angle 8 =$ _____

7. Based on your answers in problem 6, which pairs of interior and which pairs of exterior angles are equal?

8. Based on your answers in problems 6 and 7, which angles are alternate interior angles?

9. Based on your answers in problems 6 and 7, which angles are alternate exterior angles?

10. Based on your observations, alternate interior angles have _____ measure. Alternate exterior angles have _____ measure.

NAME:

Congruence
Set 1: Parallel Lines and Transversals

Station 4

At this station, you will find spaghetti noodles, a protractor, and parallel lines cut by a transversal.

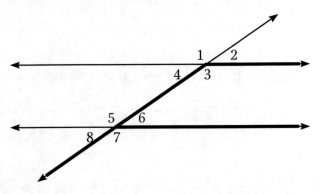

The letter "F" has been drawn on the diagram above in a bold line. Place spaghetti noodles on the "F".

1. What angles are at the inside corners of the "F"? _____

   These are called *corresponding angles*.

2. Move the spaghetti noodles around to create more "F's" that help you find the corresponding angles. Write the pairs of corresponding angles in the space below.

3. What strategy did you use to move the letter "F" around to help you find the corresponding angles?

4. Use your protractor to measure each angle to justify your answers for problem 2.

   $m\angle 1$ = _____ $m\angle 3$ = _____ $m\angle 5$ = _____ $m\angle 7$ = _____

   $m\angle 2$ = _____ $m\angle 4$ = _____ $m\angle 6$ = _____ $m\angle 8$ = _____

Congruence

Set 2: Classifying Triangles and Angle Theorems

Instruction

Goal: To provide opportunities for students to develop concepts and skills related to classifying, constructing, and describing triangles by sides and angles

Common Core Standards

Congruence

Experiment with transformations in the plane.

G-CO.1. Know precise definitions of angle, circle, perpendicular line, parallel line, and line segment, based on the undefined notions of point, line, distance along a line, and distance around a circular arc.

Prove geometric theorems.

G-CO.10. Prove theorems about triangles.

Student Activities Overview and Answer Key

Station 1

Students will use a ruler to measure side lengths of given triangles. Then they determine how to classify triangles by sides. They construct different types of triangles based on side lengths.

Answers

1. 2.5 inches, 2.5 inches, 1.5 inches
2. 1.25 inches, 1.25 inches, 1.25 inches
3. 1.75 inches, 3 inches, 2.5 inches
4. 2 sides equal; all 3 sides equal; no sides equal
5. 9.5 feet of fencing; isosceles
6. Answers will vary. Make sure all three sides are equal.
7. Answers will vary. Make sure two sides are equal.
8. Answers will vary. Make sure no sides are equal.

Congruence
Set 2: Classifying Triangles and Angle Theorems

Instruction

Station 2

Students will be given four index cards with angle measures on them, four index cards with different types of triangles, and a protractor. Students will work as a group to match the index cards with angles measures to the index cards with different types of triangles. Then they will derive how to classify triangles based on their angle measures. They will construct triangles based on their angle measures.

Answers

1. 60°, 60°, 60°, and equiangular triangle
2. 110°, 30°, 40°, and obtuse triangle
3. 90°, 45°, 45°, and right triangle
4. 35°, 80°, 65°, and acute triangle
5. One 90° angle; all angles less than 90°; one angle greater than 90°; all angles equal 60°
6. Answers will vary. Make sure all angles equal 60°.
7. Answers will vary. Make sure one angle equals 90°.
8. Answers will vary. Make sure one angle is greater than 90°.
9. Answers will vary. Make sure all angles are less than 90°.

Station 3

Students will be given graph paper, a ruler, and a protractor. Students will construct triangles given two angle measures. They will measure the third angle using the protractor. Then they will determine how to find the measure of a third angle in a triangle given the first two angles. They will use the protractor to double-check their answers.

Answers

1. Answers will vary. Make sure one angle equals 90°; 180°
2. 60°; 180°
3. Add 105° and 50°, and then subtract from 180°; 25°
4. Add 55° and 80°, and then subtract from 180°; 45°
5. Add two given angles, and then subtract from 180° to find the measure of the third angle.

Congruence
Set 2: Classifying Triangles and Angle Theorems

Instruction

Station 4

Students will be given a protractor. Students will use the protractor to measure angles in different types of triangles and exterior angles of triangles. Then they will derive a relationship between the exterior angles and the angles in the triangle.

Answers

1. 30°, 60°, 90°, $a = 90°$, $b = 150°$, and $c = 120°$
2. 125°, 35°, 20°, $a = 55°$, $b = 160°$, and $c = 145°$
3. 60°, 70°, 50°, $a = 110°$, $b = 120°$, and $c = 130°$
4. 180°
5. The exterior angle equals the sum of opposite interior angles.
6. 360°

Materials List/Setup

Station 1 ruler

Station 2 protractor; four index cards with the following sets of angles written on them:

60°, 60°, 60°   110°, 30°, 40°

90°, 45°, 45°   35°, 80°, 65°

four index cards with the following triangles on them:
(Do not label the angles, but do label the type of triangle.)

- right triangle with angle measures 90°, 45°, 45°:

   right triangle

- equiangular triangle with angle measures 60°, 60°, 60°:

   equiangular triangle

Congruence
Set 2: Classifying Triangles and Angle Theorems

Instruction

- obtuse triangle with angle measures 110°, 30°, 40°:

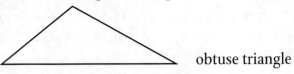
obtuse triangle

- acute angle with angle measures 35°, 80°, 65°:

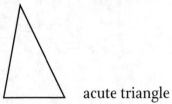
acute triangle

Station 3   graph paper; ruler; protractor
Station 4   protractor

Congruence
Set 2: Classifying Triangles and Angle Theorems

Instruction

Discussion Guide

To support students in reflecting on the activities and to gather some formative information about student learning, use the following prompts to facilitate a class discussion to "debrief" the station activities.

Prompts/Questions

1. Describe a scalene triangle.
2. Describe an isosceles triangle.
3. Describe an equilateral triangle.
4. Describe a right triangle.
5. Describe an acute triangle.
6. Describe an obtuse triangle.
7. Describe an equiangular triangle.
8. What is an exterior angle equal to in a triangle?
9. How can you find the third angle in a triangle if you are given the measure of two angles in the triangle?

Think, Pair, Share

Have students jot down their own responses to questions, then discuss with a partner (who was not in their station group), and then discuss as a whole class.

Suggested Appropriate Responses

1. In a scalene triangle, no side lengths are equal.
2. In an isosceles triangle, two side lengths are equal.
3. In an equilateral triangle, all three side lengths are equal.
4. A right triangle has one 90° angle.
5. An acute triangle has all angles less than 90°.
6. An obtuse triangle has one angle greater than 90°.
7. An equiangular triangle has all angles equal to 60°.

Congruence
Set 2: Classifying Triangles and Angle Theorems

Instruction

8. In a triangle, an exterior angle equals the sum of the opposite interior angles.

9. To find the measure of the third angle, add the two given angles, and then subtract the sum from 180°.

Possible Misunderstandings/Mistakes

- Not realizing that right triangles can have three different side lengths like scalene triangles
- Not realizing that all right triangles have one 90° angle
- Not using the correct opposite interior angles when finding the corresponding exterior angle
- Not realizing that triangles can be classified by sides and angles
- Forgetting that the sum of the angles of all triangles must equal 180°

NAME:

Congruence
Set 2: Classifying Triangles and Angle Theorems

Station 1

At this station, you will find a ruler. As a group, use the ruler to help you classify triangles by their sides. Find the length of the sides of each triangle below and record your answers on the lines.

1. Isosceles triangle

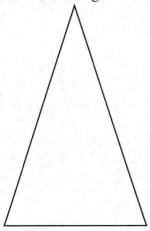

   Side lengths = _____, _____, _____

2. Equilateral triangle

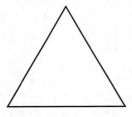

   Side lengths = _____, _____, _____

continued

Congruence
Set 2: Classifying Triangles and Angle Theorems

3. Scalene triangle

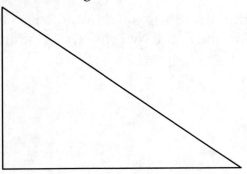

Side lengths = _____, _____, _____

4. Based on your measurements of side lengths in problems 1–3, what can you say about the relationship of the side lengths for each type of triangle?

Isosceles triangle:

Equilateral triangle:

Scalene triangle:

continued

NAME:

Congruence
Set 2: Classifying Triangles and Angle Theorems

5. Megan is going to build a triangular garden in her backyard. She sketched this diagram:

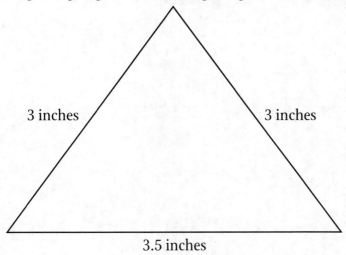

If 1 inch = 1 foot, how many feet of fencing will she need to build a fence around her garden?

_____

Her garden is an example of what type of triangle? Explain your answer.

For problems 6–8, construct an example of the given triangle. Label the length of each side.

6. Equilateral triangle:

continued

Congruence
Set 2: Classifying Triangles and Angle Theorems

7. Isosceles triangle:

8. Scalene triangle:

Congruence

Set 2: Classifying Triangles and Angle Theorems

Station 2

At this station, you will find a protractor and four index cards with the following angles written on them:

      60°, 60°, 60°   110°, 30°, 40°

      90°, 45°, 45°   35°, 80°, 65°

You will also find four index cards with illustrations of four types of triangles on them: right, acute, obtuse, and equiangular.

 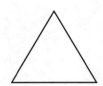 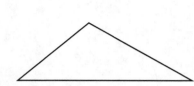

As a group, use the protractor to measure each triangle on the index cards and match it to the appropriate index card of angle measures. Write your answers on the lines below.

1. _____
2. _____
3. _____
4. _____

5. Based on your matches in problems 1–4, what can you say about the angle measures in each type of triangle?

   Right triangle: _____

   Acute triangle: _____

   Obtuse triangle: _____

   Equiangular triangle: _____

continued

Congruence
Set 2: Classifying Triangles and Angle Theorems

For problems 6–9, construct an example of the given triangle. Label the measure of each angle.

6. Equiangular triangle:

7. Right triangle:

8. Obtuse triangle:

9. Acute triangle:

NAME: _____

Congruence
Set 2: Classifying Triangles and Angle Theorems

Station 3

At this station, you will find graph paper, a ruler, and a protractor. Work as a group to construct triangles, find angle measures, and answer the questions.

1. On your graph paper, construct a right triangle that has side lengths of 6 units and 8 units and a hypotenuse of 10 units.

   Use your protractor to measure each angle in the triangle. Record your answers below.

   angle: _____; angle: _____; angle: _____

   What is the sum of the angles in the triangle? _____

2. On your graph paper, construct an equilateral triangle with side lengths of 5 units, 5 units, and 5 units.

   In an equilateral triangle all angles should measure _____.

   What is the sum of the angles in the triangle? _____.

3. On your graph paper, construct an obtuse triangle that has side lengths of your choosing. Make sure one of the angles is 105° and the second angle is 50°.

   How can you determine the measure of the third angle WITHOUT using your protractor?

   What is the measure of the third angle? _____

   Use your protractor to double-check your answer.

continued

Congruence
Set 2: Classifying Triangles and Angle Theorems

4. On your graph paper, construct an acute triangle that has side lengths of your choosing. Make sure one of the angles is 55° and the second angle is 80°.

   How can you determine the measure of the third angle WITHOUT using your protractor?

   What is the measure of the third angle? _____

   Use your protractor to double-check your answer.

5. Based on your observations in problems 1–4, how can you find the measure of the third angle of a triangle if you are given two angles in the triangle?

NAME:

Congruence
Set 2: Classifying Triangles and Angle Theorems

Station 4

At this station, you will find a protractor. Work as a group to construct each triangle, measure angles, and answer the questions.

For problems 1–3, use your protractor to measure each angle in the triangles. Then measure the exterior angles labeled *a*, *b*, and *c*.

1.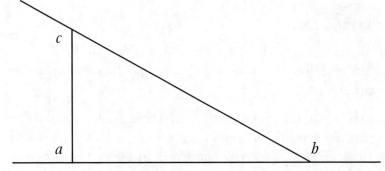

2.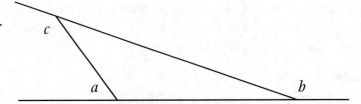

3.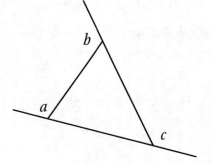

4. What is the sum of all three angles in each triangle? _____

5. What relationship does each exterior angle *a*, *b*, and *c* have to the other angles in the triangle?

6. What is the sum of all three exterior angles for each triangle? _____

Congruence

Set 3: Corresponding Parts, Transformations, and Proof

Instruction

Goal: To provide opportunities for students to develop concepts and skills related to corresponding parts, transformations, and principles of congruent triangles

Common Core Standards

Congruence

Experiment with transformations in the plane.

- **G-CO.1.** Know precise definitions of angle, circle, perpendicular line, parallel line, and line segment, based on the undefined notions of point, line, distance along a line, and distance around a circular arc.

- **G-CO.2.** Represent transformations in the plane using, e.g., transparencies and geometry software; describe transformations as functions that take points in the plane as inputs and give other points as outputs. Compare transformations that preserve distance and angle to those that do not (e.g., translation versus horizontal stretch).

- **G-CO.5.** Given a geometric figure and a rotation, reflection, or translation, draw the transformed figure using, e.g., graph paper, tracing paper, or geometry software. Specify a sequence of transformations that will carry a given figure onto another.

Understand congruence in terms of rigid motions.

- **G-CO.6.** Use geometric descriptions of rigid motions to transform figures and to predict the effect of a given rigid motion on a given figure; given two figures, use the definition of congruence in terms of rigid motions to decide if they are congruent.

- **G-CO.7.** Use the definition of congruence in terms of rigid motions to show that two triangles are congruent if and only if corresponding pairs of sides and corresponding pairs of angles are congruent.

- **G-CO.8.** Explain how the criteria for triangle congruence (ASA, SAS, and SSS) follow from the definition of congruence in terms of rigid motions.

Expressing Geometric Properties with Equations

Use coordinates to prove simple geometric theorems algebraically.

- **G-GPE.4.** Use coordinates to prove simple geometric theorems algebraically.

Congruence
Set 3: Corresponding Parts, Transformations, and Proof

Instruction

Student Activities Overview and Answer Key

Station 1

Students will use a ruler and protractor to prove that two triangles and corresponding parts are congruent. Then they will use the properties of triangles to prove that two triangles and corresponding parts are congruent.

Answers

1. Measure lengths of corresponding sides and see if they are equal.
2. Measure corresponding angles and see if they are equal.
3.

$\overline{DB} \perp \overline{AC}$	Given
$\angle DBA$ and $\angle DBC$ are right angles	Definition of perpendicular lines
$\triangle DBA$ and $\triangle DBC$ are right triangles.	A right triangle has a right angle.
$\overline{DB} \cong \overline{DB}$	Reflexive property of congruence
$\triangle DBA \cong \triangle DBC$	Hypotenuse leg postulate
$\overline{AB} \cong \overline{AC}$	CPCTC (Corresponding parts of congruent triangles are congruent)

4. Answers will vary. Possible answer: designing a skyscraper
5. Yes, corresponding parts of congruent triangles are congruent.

Station 2

Students will be given graph paper, a ruler, push pins, and rubber bands. Students will construct triangles and perform translations and dilations. They will show how to translate congruent triangles. They will explain why dilations of triangles do not yield congruent triangles.

Answers

1. Triangle with vertices (–7, –7), (7, –7), and (7, 9)
2. Triangle with vertices (–8, –2), (6, –2), and (6, 14); yes; n/a
3. Triangles with vertices (–14, 4), (7, 4), and (7, 20); no, sides are different lengths; answers may vary. Sample answer: (–7, 4), (7, 4), and (7, 20)
4. Triangle with vertices (0, –5), (8, –5), and (0, 5)
5. (0, –10), (16, –10), (0, 10)
6. No, because dilation changed the size of the triangle.

Congruence
Set 3: Corresponding Parts, Transformations, and Proof

Instruction

Station 3

Students will be given graph paper, a ruler, and a cardboard triangle. Students will perform reflections and identify congruent triangles that have been reflected. Then they will use the cardboard triangle to perform a rotation and explain how it relates to congruent triangles.

Answers

1. (–4, 2), (–7, 2), (–7, 7); (–4, –2), (–7, –2); (–7, –7). Yes, because the size and shape remained the same.
2. A and B
3. Yes, because the size and shape remained the same.
4. They have the same size and shape.

Station 4

Students will be given four index cards with the following written on them: SSS; SAS; ASA; AAS. Students will work together to match the index cards to real-world examples of SSS, SAS, ASA, and AAS. Then they will explain how SSS, SAS, ASA, and AAS relate to congruent triangles.

Answers

1. ASA
2. AAS
3. SSS
4. SAS
5. Answers will vary.
6. side-side-side; side-angle-side; angle-side-angle; angle-angle-side. These are ways to prove two triangles are congruent.

Materials List/Setup

Station 1	ruler; protractor
Station 2	graph paper; ruler; push pins; rubber bands
Station 3	graph paper; ruler; cardboard triangle created from a triangle with vertices (4, 4), (10, 4), and (6, 12) in the coordinate plane
Station 4	four index cards with the following written on them: SSS; SAS; ASA; AAS

Congruence
Set 3: Corresponding Parts, Transformations, and Proof

Instruction

Discussion Guide

To support students in reflecting on the activities and to gather some formative information about student learning, use the following prompts to facilitate a class discussion to "debrief" the station activities.

Prompts/Questions

1. If two triangles are congruent, then what parts of the triangles are congruent?
2. Does a translation create a congruent triangle? Why or why not?
3. Does dilation create a congruent triangle? Why or why not?
4. Does a rotation create a congruent triangle? Why or why not?
5. Does a reflection create a congruent triangle? Why or why not?
6. What are four ways to prove two triangles are congruent if you know only three pieces of information about the triangles?

Think, Pair, Share

Have students jot down their own responses to questions, then discuss with a partner (who was not in their station group), and then discuss as a whole class.

Suggested Appropriate Responses

1. Corresponding parts of congruent triangles are congruent.
2. Yes, because the size and shape of the triangle remain the same.
3. No, because the size does not remain the same.
4. Yes, because the size and shape of the triangle remain the same.
5. Yes, because the size and shape of the triangle remain the same.
6. SSS, ASA, AAS, SAS

Possible Misunderstandings/Mistakes

- Not reflecting, rotating, or translating a triangle correctly to create a congruent triangle
- Not realizing that dilation changes the size of the triangle
- Not identifying the relationship between the locations of the sides and angles in triangles when selecting ASA, AAS, or SAS to prove two triangles are congruent
- Incorrectly identifying corresponding parts of triangles

Congruence
Set 3: Corresponding Parts, Transformations, and Proof

Station 1

At this station, you will find a ruler and a protractor. Work as a group to answer the questions.

Nigel is building a doghouse. He sketched this diagram of the roof:

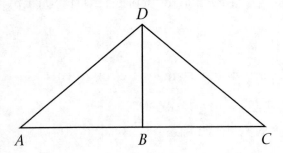

It is important that the roof is symmetrical. Therefore, $\triangle DBA \cong \triangle DBC$.

1. How can you use your ruler to show that $\triangle DBA \cong \triangle DBC$ and $\overline{AB} \cong \overline{BC}$?

2. How can you use your protractor to show that $\triangle DBA \cong \triangle DBC$ and $\overline{AB} \cong \overline{BC}$?

3. What if you didn't have a ruler or protractor? How could you use the angle and side properties of triangles to prove that $\triangle DBA \cong \triangle DBC$ and $\overline{AB} \cong \overline{BC}$?

   For example, if it is given that $\overline{DC} \cong \overline{DA}$ and $\overline{DB} \perp \overline{AC}$, how can you prove that $\triangle DBA \cong \triangle DBC$ and $\overline{AB} \cong \overline{BC}$?

continued

Congruence
Set 3: Corresponding Parts, Transformations, and Proof

4. Provide an example of a real-world situation in which proving two triangles are congruent using the method in problem 3 is much more feasible than actually measuring the two triangles. Explain your reasoning.

5. If you know that two triangles are congruent, are all corresponding sides and angles congruent? Why or why not?

Congruence
Set 3: Corresponding Parts, Transformations, and Proof

Station 2

At this station, you will find graph paper, a ruler, push pins, cardboard, and rubber bands. Mount the graph paper on cardboard. Create an *x*- and *y*-axis on your graph paper.

1. As a group, construct a triangle on your graph paper by placing push pins at the points (–7, –7), (7, –7), and (7, 9). Place a rubber band around pairs of push pins to create a triangle.

2. Construct a second triangle on your graph paper by placing push pins at the points (–8, –2), (6, –2), and (6, 14). Place a rubber band around pairs of push pins to create a triangle.

    Is this triangle congruent to the triangle in problem 1? Why or why not?

    If the two triangles are not congruent, how can you modify the second triangle so it is congruent to the first triangle?

3. Construct a third triangle on your graph paper by placing push pins at the points (–14, 4), (7, 4), and (7, 20). Place a rubber band around pairs of push pins to create a triangle.

    Is this triangle congruent to the triangle in problem 1? Why or why not?

Congruence
Set 3: Corresponding Parts, Transformations, and Proof

If the two triangles are not congruent, how can you modify the third triangle so it is congruent to the first triangle?

Remove all the push pins and rubber bands from the graph.

4. On your graph paper, construct a triangle by placing push pins at the points (0, −5), (8, −5), and (0, 5). Place one rubber band around all of the push pins to create a triangle.

5. Construct a second triangle by dilating the first triangle to twice its size about the origin, (0, 0).

   At what points did you place the push pins for this new triangle?

6. Are the triangles you created in problems 4 and 5 congruent? Why or why not?

Congruence
Set 3: Corresponding Parts, Transformations, and Proof

Station 3

At this station, you will find graph paper, a ruler, and a cardboard triangle.

As a group, create an *x*- and *y*-axis on your graph paper and work together to answer the questions.

1. On your graph paper, construct a triangle that has vertices (4, 2), (7, 2), and (7, 7).

   Reflect this triangle across the *y*-axis. What are the vertices of this new triangle?

   Reflect this second triangle across the *x*-axis. What are the vertices of this new triangle?

   Are the three triangles congruent? Why or why not?

Use the following diagram to answer question 2.

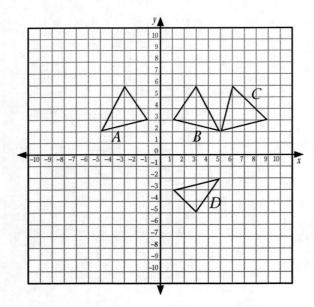

2. Which triangles are congruent? Explain your answer.

continued

Congruence

Set 3: Corresponding Parts, Transformations, and Proof

3. On a new graph, place the cardboard triangle so its vertices are at (4, 4), (10, 4), and (6, 12). Trace the cardboard triangle.

   Rotate the cardboard triangle 90° about the point (4, 4). Trace the cardboard triangle. Are the triangles congruent?

4. Based on your observations in problems 1–3, when properly reflecting or rotating triangles, they are congruent if they _____

Congruence
Set 3: Corresponding Parts, Transformations, and Proof

Station 4

At this station, you will find four index cards with the following written on them:

SSS; SAS; ASA; AAS

Work as a group to match each index card to the following real-world situations described.

1. A stained glass window contains two triangles.

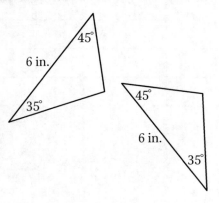

   Matching index card: _____

2. Anna creates earrings in the shape of scalene triangles.

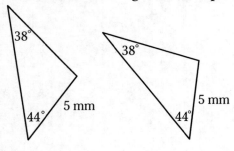

   Matching index card: _____

continued

NAME:

Congruence
Set 3: Corresponding Parts, Transformations, and Proof

3. Peter designed a logo for his Web design company. Each triangle has side lengths of 3 inches, 3 inches, and 3 inches.

   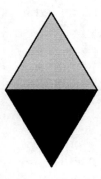

   Matching index card: _____

4. Janice designed two new buttons for her Web site. One button was a purple isosceles triangle with a 30° angle between the congruent sides. The other button was a green isosceles triangle with a 30° angle between the congruent sides.

   Matching index card: _____

5. What strategy did you use to match the cards?

6. Based on your observations in problems 1–4, what do you think each index card stands for as far as congruent triangles?

   SSS = _____

   SAS = _____

   ASA = _____

   AAS = _____

Congruence

Set 4: Special Congruent Triangles

Instruction

Goal: To provide opportunities for students to develop concepts and skills related to proving right, isosceles, and equilateral triangles congruent using real-world problems

Common Core Standards

Congruence

Experiment with transformations in the plane.

G-CO.1. Know precise definitions of angle, circle, perpendicular line, parallel line, and line segment, based on the undefined notions of point, line, distance along a line, and distance around a circular arc

Understand congruence in terms of rigid motions.

G-CO.7. Use the definition of congruence in terms of rigid motions to show that two triangles are congruent if and only if corresponding pairs of sides and corresponding pairs of angles are congruent.

Similarity, Right Triangles, and Trigonometry

Define trigonometric ratios and solve problems involving right triangles.

G-SRT.8. Use trigonometric ratios and the Pythagorean theorem to solve right triangles in applied problems.★

Expressing Geometric Properties with Equations

Use coordinates to prove simple geometric theorems algebraically.

G-GPE.4. Use coordinates to prove simple geometric theorems algebraically.

Student Activities Overview and Answer Key

Station 1

Students will be given graph paper and a ruler. They construct a rectangle, turn it into two right triangles, and examine the relationships between the hypotenuse and legs. Then they construct right triangles and prove two right triangles are congruent if their hypotenuse and corresponding leg are equal.

Answers

1. two right triangles; hypotenuse and length of legs; yes
2. leg

Congruence
Set 4: Special Congruent Triangles

Instruction

3. Pythagorean theorem

4. Use the Pythagorean theorem to find the length of the third leg. Use the hypotenuse and corresponding leg to prove right triangles are congruent.

5. Corresponding legs = 6 or 8 and hypotenuse = 10. Therefore, right triangles are congruent.

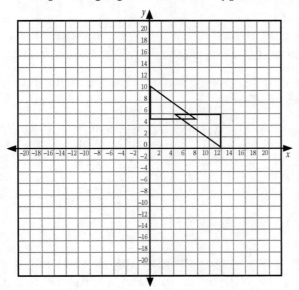

Station 2

Students will be given a ruler and a protractor. They will prove two triangles are congruent using the properties of congruent and isosceles triangles. Then they will justify their answer using the ruler and protractor.

Answers

1. two

2. opposite the equal sides

3. $\overline{BA} \cong \overline{BC}$

4. $\triangle ABC$ is isosceles and $\overline{BA} \cong \overline{BC}$, so $\angle BAD \cong \angle BCE$.

   Or, since $\overline{AD} \cong \overline{DB}$, $\triangle ADB$ is isosceles, and $\angle BAD \cong \angle ABD$. Since $\overline{EB} \cong \overline{EC}$, $\triangle BEC$ is isosceles, and $\angle CBE \cong \angle BCE$. Therefore, $\triangle ADB$ and $\triangle BEC$ have corresponding congruent sides of AB and BC. $\triangle ADB \cong \triangle BEC$.

5. Answers will vary.

Congruence
Set 4: Special Congruent Triangles

Instruction

Station 3

Students will be given tracing paper and scissors. They will use the tracing paper and scissors to determine if triangles are equilateral and congruent. Then they will use the properties of congruent and equilateral triangles to prove equilateral triangles are congruent.

Answers

1. Answers will vary. Possible answer: Trace the triangles, cut them out, and place them on top of each other to see if they are the same size and shape; yes, the gardens are of equal size.

2. All the angles equal 60°; since the large triangle is equilateral and has side lengths of 10 feet, then all of the triangles have side lengths of 5 feet.

3. Answers will vary. Possible answer: Use properties when you create scale models and/or can't physically measure the figure.

4. 60°, angles, side lengths

Station 4

Students will be given graph paper and a ruler. They will use the graph paper and ruler to model real-world examples of right triangles. They will use the vertices and properties of the triangles to find congruent right triangles.

Answers

1. (0, 0), (0, 3), (–4, 3)

2. (0, 0), (0, –4), (3, 0)

3. Yes, because they have the same shape and size.

4. Answers will vary. Make sure the hypotenuse is 3.6 units and the legs are 2 and 3 units.

5. Answers will vary. One possible solution: Walk northwest 3.6 miles, walk south 3 miles, and walk east 2 miles. The sides and angles must be congruent to the triangle that they walked on Friday.

Materials List/Setup

Station 1	graph paper; ruler
Station 2	ruler; protractor
Station 3	tracing paper; scissors
Station 4	graph paper; ruler

Congruence
Set 4: Special Congruent Triangles

Instruction

Discussion Guide

To support students in reflecting on the activities and to gather some formative information about student learning, use the following prompts to facilitate a class discussion to "debrief" the station activities.

Prompts/Questions
1. What is the HL congruence theorem for right triangles?
2. How many congruent sides does an isosceles triangle have?
3. Which angles in an isosceles triangle are congruent?
4. How many sides in an equilateral triangle are congruent?
5. How many angles in an equilateral triangle are congruent?

Think, Pair, Share

Have students jot down their own responses to questions, then discuss with a partner (who was not in their station group), and then discuss as a whole class.

Suggested Appropriate Responses
1. Two right triangles are congruent if their hypotenuse and a corresponding leg are congruent.
2. two
3. angles opposite the two congruent sides
4. three
5. three

Possible Misunderstandings/Mistakes
- Incorrectly identifying corresponding legs when using hypotenuse-leg congruence for right triangles
- Not understanding that equilateral triangles are also equiangular and vice versa
- Not realizing that congruent angles in an isosceles triangle are opposite the congruent sides

Congruence
Set 4: Special Congruent Triangles

Station 1

At this station, you will find graph paper and a ruler. As a group, create an *x*- and *y*-axis on your graph paper.

1. On your graph paper, graph a rectangle that has vertices (2, 2), (2, 6), (5, 2), and (5, 6). Construct a diagonal in the rectangle.

   What two shapes have you created? _____

   What do the two shapes have in common?

   Are the shapes congruent? Explain your answer.

2. Graph a right triangle with vertices (–1, 1), (–7, 1), and (–1, 9). Graph a second right triangle with vertices (–1, 1), (–1, 9), and (5, 1).

   What do both of these triangles have in common? Explain your answer.

3. A right triangle has a leg length of 3 and a hypotenuse of 5. How can you find the length of the other leg?

   This means that given the hypotenuse and leg of a right triangle you can always find the length of the third leg.

continued

Congruence
Set 4: Special Congruent Triangles

4. Based on your observations in problems 1–3, how can you prove that right triangles are congruent if the hypotenuse and one corresponding leg are equal in both triangles?

5. Use hypotenuse-leg congruence of right triangles to prove that a triangle with vertices (0, 5), (8, 5), and (0, 11) is congruent to a triangle with vertices (12, 0), (12, 6), and (4, 6).

   Construct these triangles on your graph paper. Then show your work and answer in the space below.

Congruence
Set 4: Special Congruent Triangles

Station 2

At this station, you will find a ruler and a protractor. Work as a group to solve the following real-world example of isosceles and congruent triangles:

Darla is working on a logo design for her blog. Here's what she has so far:

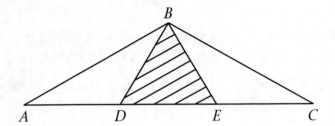

1. Isosceles triangles have how many equal sides? _____

2. Isosceles triangles have two equal angles. Where are these equal angles located?

3. If △ABC is an isosceles triangle, then which sides of the triangle are congruent?

   Points D and E are on $\overline{AC}$ such that $\overline{AD} \cong \overline{DB}$ and $\overline{BE} \cong \overline{EC}$.

4. How can you show that △ADB ≅ △BEC using properties of isosceles and congruent triangles? Show your work and answer in the space below.

5. Use your ruler and protractor to prove that △ADB ≅ △BEC. Explain how to do this.

NAME:

Congruence
Set 4: Special Congruent Triangles

Station 3

At this station, you will find tracing paper and scissors. Read the problem scenario below, then work as a group to answer the questions.

The figure below is a courtyard between three apartment buildings in a city. Each building owner owns one of the outer equilateral triangles, which they plan to turn into gardens. The middle equilateral triangle contains a fountain.

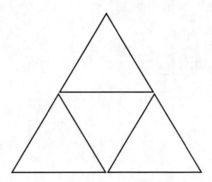

1. The building owners want to make sure that they each have the same size garden. How can you use this figure, tracing paper, and scissors to determine if the gardens are of equal size?

   Are the gardens equal in size? Explain your answer.

2. The entire courtyard is an equilateral triangle with side lengths of 10 feet. The middle triangle, containing the fountain, is an equilateral triangle with vertices at the midpoint of each side of the entire courtyard.

   How can you use the properties of equilateral and congruent triangles to prove that the gardens are of equal size?

continued

Congruence
Set 4: Special Congruent Triangles

3. Why is it useful to know the properties of congruent triangles when designing gardens and other architectural projects?

4. If an equilateral triangle is equiangular, what is the measure of each angle in the triangle?
   _____

   Therefore, congruent equilateral triangles have equal _____ and equal
   _____.

Congruence
Set 4: Special Congruent Triangles

Station 4

At this station, you will find graph paper and a ruler. Work as a group to answer the questions.

Karl and Simone walk their dogs in the park. Each day they start at the same spot in the center of the park. They walk different trails in the park on different days:

- On Mondays, they walk 3 miles to the north, then 4 miles to the west, and then 5 miles to the southeast.
- On Wednesdays, they walk 4 miles to the south, 5 miles to the northeast, and 3 miles to the west.
- On Fridays, they walk 3.6 miles to the northeast, 3 miles south, and 2 miles west.

1. On your graph paper, plot a point at (0, 0) that represents the center of the park. Construct a triangle that represents the path they walk on Mondays.

   What are the vertices of this triangle?

2. On the same graph, construct a triangle that represents the path they walk on Wednesdays.

   What are the vertices of this triangle?

3. Are the paths they walk on Mondays and Wednesdays congruent right triangles? Why or why not?

4. On your graph paper, construct a triangle that represents the path they walk on Fridays.

5. Karl and Simone also want to walk their dogs on Saturday. They want to walk a triangle that is congruent to the triangle they walk on Fridays.

   What is a possible walking path (including directions and miles) they could walk on Saturday? Explain your answer.

Congruence

Set 5: Bisectors, Medians, and Altitudes

Instruction

Goal: To provide opportunities for students to develop concepts and skills related to identifying and constructing angle bisectors, perpendicular bisectors, medians, altitudes, incenters, circumcenters, centroids, and orthocenters

Common Core Standards

Congruence

Experiment with transformations in the plane.

G-CO.1. Know precise definitions of angle, circle, perpendicular line, parallel line, and line segment, based on the undefined notions of point, line, distance along a line, and distance around a circular arc.

Prove geometric theorems.

G-CO.10. Prove theorems about triangles.

Make geometric constructions.

G-CO.12. Make formal geometric constructions with a variety of tools and methods (compass and straightedge, string, reflective devices, paper folding, dynamic geometric software, etc.).

Student Activities Overview and Answer Key

Station 1

Students will be given a ruler and a compass. Students will construct and identify angle bisectors. They will derive the ratio between the two segments created by the angle bisector and the other sides of the triangle. Then they will construct and find the incenter based on the angle bisectors and construct a circle inscribed in the triangle.

Congruence
Set 5: Bisectors, Medians, and Altitudes

Instruction

Answers

1.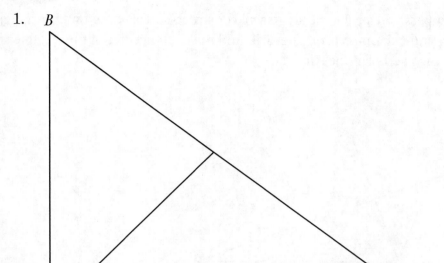

2. about 2.15 in., about 2.85 in., 3 in., 4 in.

3. $BD/DC = AB/AC$; proportion

4. Triangle sizes will vary.

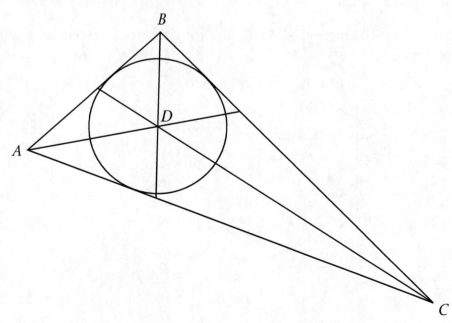

5. No, because only one circle that has a center equidistant from each side of the triangle can be inscribed in the triangle.

Congruence
Set 5: Bisectors, Medians, and Altitudes

Instruction

Station 2

Students will be given a compass and a ruler. Students will construct isosceles and obtuse triangles. They will construct the perpendicular bisectors. They will find the circumcenter of the triangle and examine its relationship to each vertex of the triangle.

Answers

1.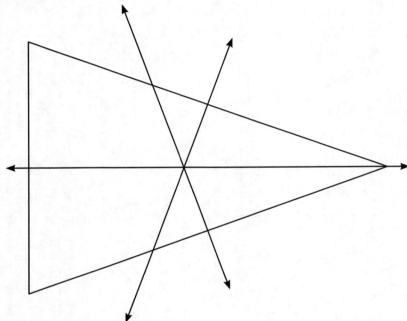

2. interior

3. Answers will vary, but the distances should be equal; the circumcenter is equidistant from all vertices of the triangle.

4–6.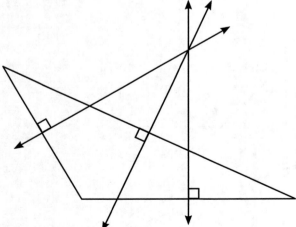

7. exterior

8. Answers will vary, but the distances should be equal; the circumcenter is equidistant from all vertices of the triangle.

Congruence
Set 5: Bisectors, Medians, and Altitudes

Instruction

Station 3

Students will be given a compass and ruler. Students will construct a triangle. Students will construct the medians and centroid of a triangle. Then they will measure the side lengths of the triangle, medians, and distance from the centroid to each vertex. They will derive the relationship between the centroid, medians, and vertices.

Answers

1.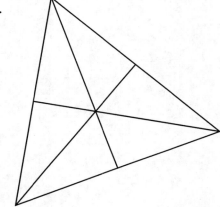

2. Answers will vary.
3. Answers will vary.
4. Answers will vary.
5. The centroid is twice as far from a given vertex than it is from the point of intersection of the median to the opposite side from that vertex.
6. No, the centroid is always on the interior of the triangle. By definition, medians are between each vertex and the midpoint of each side. Therefore, the centroid must always be inside the triangle.

Station 4

Students will be given a compass and a ruler. Students will construct altitudes and orthocenters for acute, obtuse, and right triangles. They will analyze the length of altitudes and locations of orthocenters.

Congruence
Set 5: Bisectors, Medians, and Altitudes

Instruction

Answers

1.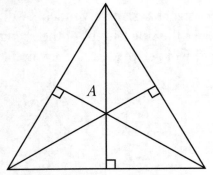

2. Answers will vary.

3. inside the triangle

4.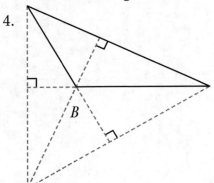

5. Answers will vary.

6. outside the triangle

7.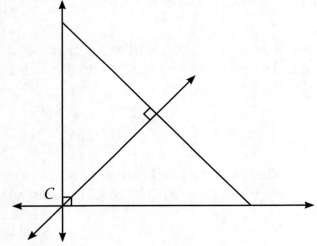

8. Answers will vary.

9. on the vertex of the 90° angle

Congruence
Set 5: Bisectors, Medians, and Altitudes

Instruction

Materials List/Setup

Station 1   compass; ruler
Station 2   compass; ruler
Station 3   compass; ruler
Station 4   compass; ruler

Congruence
Set 5: Bisectors, Medians, and Altitudes

Instruction

Discussion Guide

To support students in reflecting on the activities and to gather some formative information about student learning, use the following prompts to facilitate a class discussion to "debrief" the station activities.

Prompts/Questions
1. What is the name of the intersection of the angle bisectors?
2. What is the name of the intersection of the perpendicular bisectors?
3. What is a median of a triangle?
4. What is the name of the intersection of the medians of a triangle?
5. What is an altitude of a triangle?
6. What is the name of the intersection of the altitudes of a triangle?

Think, Pair, Share

Have students jot down their own responses to questions, then discuss with a partner (who was not in their station group), and then discuss as a whole class.

Suggested Appropriate Responses
1. incenter
2. circumcenter
3. The median of a triangle is a line segment drawn from a vertex to the midpoint of the opposite side.
4. centroid
5. The altitude of a triangle is the perpendicular distance from a vertex of a figure to the side opposite the vertex.
6. orthocenter

Possible Misunderstandings/Mistakes
- Confusing how to identify angle bisectors versus perpendicular bisectors
- Confusing how to identify and construct medians versus altitudes
- Mixing up the terms incenter, circumcenter, centroid, and orthocenter

Congruence
Set 5: Bisectors, Medians, and Altitudes

Station 1

At this station, you will find a ruler and a compass. Work as a group to answer the questions.

1. In the space below, construct a right triangle *ABC* with legs equal to 3 inches and 4 inches, and a hypotenuse of 5 inches. Label the right angle as *A*.

   Use the compass to bisect the triangle at vertex *A*. Name the point where the bisector intersects $\overline{BC}$ as point *D*.

2. What is the length of $\overline{BD}$? _____

   What is the length of $\overline{DC}$? _____

   What is the length of $\overline{AB}$? _____

   What is the length of $\overline{AC}$? _____

3. Based on your observations in problems 1–2, how do $\overline{BD}$ and $\overline{DC}$ relate to $\overline{AB}$ and $\overline{AC}$? Show your work and answer in the space below.

   Complete this statement: An angle bisector divides the side opposite the bisected angle into two segments that are the same _____ as the other sides of the triangle.

continued

Congruence

Set 5: Bisectors, Medians, and Altitudes

4. In the space below, construct an acute triangle with angles measuring 40°, 80°, and 60°.

    Bisect each angle.

5. The point where the angle bisectors meet is called the **incenter**. On the triangle you constructed for problem 4, use the compass to construct a circle, with the incenter as its center point.

    Can you inscribe another circle within the triangle that is also equidistant from each side of the triangle? Why or why not?

Congruence
Set 5: Bisectors, Medians, and Altitudes

Station 2

At this station, you will find a compass and a ruler. Work as a group to answer the questions.

1. In the space below, construct an isosceles triangle with angles measuring 70°, 70°, and 40°.

   Find the midpoint of each side of the triangle.

   Construct a perpendicular bisector at the midpoint of each side of the triangle.

2. Do the three perpendicular bisectors meet at a point that is on the interior or exterior of your circle? _____

   The point of intersection of the perpendicular bisectors is called the **circumcenter**.

3. What is the distance from the circumcenter to each vertex in the triangle?

   _____
   _____
   _____

   Based on your measurements, what can you say about the relationship between the circumcenter and each vertex of the triangle?

continued

Congruence
Set 5: Bisectors, Medians, and Altitudes

4. In the space below, construct an obtuse triangle with angles measuring 120°, 35°, and 25°.

5. Find the midpoint of each side of the triangle.

6. Construct a perpendicular bisector at the midpoint of each side of the triangle.

7. Do the three perpendicular bisectors meet at a point that is on the interior or exterior of your circle? _____

8. What is the distance from the circumcenter to each vertex in the triangle?
   _____
   _____
   _____

   Based on your measurements, what can you say about the relationship between the circumcenter and each vertex of the triangle?

Congruence

Set 5: Bisectors, Medians, and Altitudes

Station 3

At this station, you will find a ruler and a compass. Work as a group to answer the questions.

1. In the space below, construct an acute triangle *ABC*.

   Find the midpoint of each side of the triangle.

   Construct a median by drawing a straight line from each midpoint to the opposite vertex.

   Label the intersection of the medians as point *C*. This point is called the **centroid** of the triangle.

2. What is the length of each side of the triangle you have created?

   _____
   _____
   _____

3. What is the length of each median?

   _____
   _____
   _____

4. What is the length between the centroid and each vertex?

   _____
   _____
   _____

continued

Congruence
Set 5: Bisectors, Medians, and Altitudes

5. Based on your observations in problems 1–4, what can you say about the relationship between the centroid and a given vertex and the centroid and the point of intersection of the median to the opposite side from that vertex?

6. Can the centroid ever occur on the outside of a triangle? Why or why not?

Congruence

Set 5: Bisectors, Medians, and Altitudes

Station 4

At this station, you will find a ruler and a compass. Work as a group to answer the questions.

1. In the space below, construct an acute triangle.

    Construct an altitude from each vertex.

    Label the intersection of the altitudes as point A. This intersection is called the **orthocenter**.

2. What is the length of each altitude?

    _____

    _____

    _____

3. Is the orthocenter on the inside, on the outside, or on a vertex of the triangle?

    _____

4. In the space below, construct an obtuse triangle.

    Construct an altitude from each vertex.

    Label the orthocenter as B.

continued

Congruence
Set 5: Bisectors, Medians, and Altitudes

5. What is the length of each altitude?

   _____

   _____

   _____

6. Is the orthocenter on the inside, on the outside, or on a vertex of the triangle?

   _____

7. In the space below, construct a right triangle.

   Construct an altitude from each vertex.
   Label the orthocenter as *C*.

8. What is the length of each altitude?

   _____

   _____

   _____

9. Is the orthocenter on the inside, on the outside, or on a vertex of the triangle?

   _____

Congruence

Set 6: Triangle Inequalities

Instruction

Goal: To provide opportunities for students to develop concepts and skills related to inequality theorems for triangles and the hinge theorem

Common Core Standards

Congruence

Experiment with transformations in the plane.

- **G-CO.1.** Know precise definitions of angle, circle, perpendicular line, parallel line, and line segment, based on the undefined notions of point, line, distance along a line, and distance around a circular arc.

Prove geometric theorems.

- **G-CO.10.** Prove theorems about triangles.

Student Activities Overview and Answer Key

Station 1

Students will be given a ruler. Students will examine the relationship between the sides of a triangle. They will realize the sum of two sides of a triangle is greater than the third side. They will use their ruler to justify their answer. Then they will find possible values for a third side of a triangle given two sides of the triangle.

Answers

1. The sum of the lengths of any two sides of a triangle must be greater than the third side.
2. $0 < x < 6$
3. Yes, because $x = 2$ inches.
4. $14 < x < 34$
5. yes
6. no
7. no
8. yes

Congruence
Set 6: Triangle Inequalities

Instruction

Station 2

Students will be given graph paper, a protractor, and a ruler. Students will construct triangles and measure the angles and side lengths of the triangles. Then they will derive a relationship between the angles and side lengths of the triangles based on their observations.

Answers

1. Answers will vary; answers will vary; 90°
2. Answers will vary; answers will vary; answers will vary.
3. 30°, 46°, 104°; about 2 in., about 2.75 in., about 3.75 in.
4. In a triangle, the longest side is across from the largest angle.

Station 3

Students will be given graph paper, a protractor, and a ruler. Students will work together to construct equilateral, obtuse, acute, and right triangles. They will find the exterior angles for each triangle. Then they will derive the relationship between exterior angles and nonadjacent angles in triangles.

Answers

1. 120°, 120°, 120°; exterior angles are greater than the measure of each nonadjacent interior angle.
2. 82°, 149°, 129°; exterior angles are greater than the measure of each nonadjacent interior angle.
3. 137°, 118°, 105°; exterior angles are greater than the measure of each nonadjacent interior angle.
4. 135°, 135°, 90°; exterior angles are greater than the measure of each nonadjacent interior angle.
5. In triangles, exterior angles are greater than the measure of each nonadjacent interior angle.

Station 4

Students will be given spaghetti noodles, graph paper, a protractor, and a ruler. Students will use the spaghetti noodles to create congruent triangles. They will measure the length of the third side and the included angle for each triangle. Then they will derive the hinge theorem for triangles.

Answers

1. Answers will vary.
2. Yes; answers will vary.

Congruence
Set 6: Triangle Inequalities

Instruction

3. If two sides of one triangle are congruent to two sides of another triangle, and the included angle of the first triangle is larger than the included angle of the second triangle, then the third side of the first triangle is longer than the third side of the second triangle.

4. Answers will vary.

5. Answers will vary; less than

6. If two sides of one triangle are congruent to two sides of another triangle, and the included angle of the first triangle is larger than the included angle of the second triangle, then the third side of the first triangle is longer than the third side of the second triangle.

7. Answers will vary.

Materials List/Setup

Station 1 ruler

Station 2 graph paper; protractor; ruler

Station 3 graph paper; protractor; ruler

Station 4 six dry spaghetti noodles that are 1 inch, 2 inches, 2.5 inches, 3.5 inches, 4 inches, and 5 inches in length; graph paper; protractor; ruler

Congruence
Set 6: Triangle Inequalities

Instruction

Discussion Guide

To support students in reflecting on the activities and to gather some formative information about student learning, use the following prompts to facilitate a class discussion to "debrief" the station activities.

Prompts/Questions

1. What is the relationship between the three side lengths in a triangle?
2. What is the relationship between the measure of the angles of a triangle and length of the opposite sides?
3. What is the relationship between the exterior angle of a triangle and the two nonadjacent interior angles of the triangle?
4. What is the hinge theorem?

Think, Pair, Share

Have students jot down their own responses to questions, then discuss with a partner (who was not in their station group), and then discuss as a whole class.

Suggested Appropriate Responses

1. The sum of any two side lengths of a triangle must be greater than the length of the third side.
2. In a triangle, the largest angle is opposite the longest side.
3. The exterior angle is greater than the measure of either nonadjacent interior angle.
4. If two sides of one triangle are congruent to two sides of another triangle, and the included angle of the first triangle is larger than the included angle of the second triangle, then the third side of the first triangle is longer than the third side of the second triangle.

Possible Misunderstandings/Mistakes

- Not finding a range of values for a third side length when given two sides of the triangle
- Not recognizing that the sum of two sides of a triangle must be greater than the third side
- Not recognizing that the largest angle measure in a triangle is opposite the longest side
- Measuring the wrong angles and/or sides when comparing triangles using the hinge theorem

NAME: _____

Congruence
Set 6: Triangle Inequalities

Station 1

At this station, you will find a ruler. Work as a group to answer the questions.

1. For each triangle below, what is the relationship between the sum of two sides of the triangle to the third side? Show your work in the space next to each triangle.

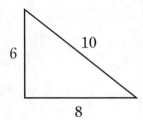

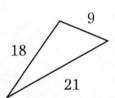

2. In the isosceles triangle below, what are possible values for the third side, x?

   Hint: Use your observations from problem 1. The possible values of x will be in an inequality statement in the form ____ < x < _____.

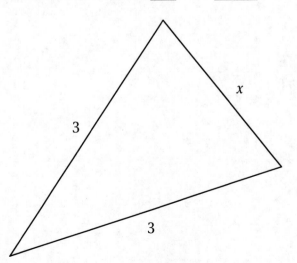

continued

Congruence
Set 6: Triangle Inequalities

3. Use your ruler to measure the triangle. Does this measurement fall into the range of possible values for *x* you found? _____

4. A triangle has side lengths of 10 ft and 24 ft. What are the possible lengths of the third side? Show your work and answer in the space below.

For problems 5–8, determine whether the triangle is possible. Explain your answer.

5. Triangle with sides of 5 in., 7 in., and 9 in.

6. Triangle with sides of 1 cm, 3 cm, and 1 cm

7. Triangle with sides of 0.5 m, 0.25 m, and 0.25 m

8. Triangle with sides of 14 ft, 17 ft, and 20 ft

Congruence
Set 6: Triangle Inequalities

Station 2

At this station, you will find graph paper, a protractor, and a ruler. Work as a group to construct the triangles and answer the questions.

1. On your graph paper, graph a right triangle that has legs that are 3 units and 4 units, and a hypotenuse of 5 units.

   What is the measure of the angle across from the 3-unit leg? _____

   What is the measure of the angle across from the 4-unit leg? _____

   What is the measure of the angle across from the 5-unit hypotenuse? _____

2. On your graph paper, graph an obtuse triangle with angle measures of 110°, 30°, and 40°.

   What is the length of the side across from the 110° angle? _____

   What is the length of the side across from the 30° angle? _____

   What is the length of the side across from the 40° angle? _____

continued

NAME:

Congruence
Set 6: Triangle Inequalities

3. Use your protractor and ruler to measure the triangle below.

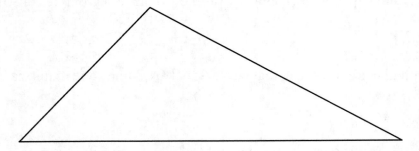

What is the measure of each angle? _____

What is the measure of each side? _____

4. Based on your observations in problems 1–4, what is the relationship between the measure of the angle and the length of the opposite side? Explain your answer.

Congruence
Set 6: Triangle Inequalities

Station 3

At this station, you will find graph paper, a protractor, and a ruler. Work as a group to construct the triangles and answer the questions.

1. On your graph paper, construct an equilateral triangle.

   What is the measure of each exterior angle? _____

   What is the relationship between each exterior angle and the measure of each nonadjacent interior angle?

2. On your graph paper, construct an obtuse triangle with angle measures of 98°, 31°, and 51°.

   What is the measure of each exterior angle? _____

   What is the relationship between each exterior angle and the measure of each nonadjacent interior angle?

3. On your graph paper, construct an acute triangle with angle measures of 43°, 62°, and 75°.

   What is the measure of each exterior angle? _____

   What is the relationship between each exterior angle and the measure of each nonadjacent interior angle?

continued

Congruence
Set 6: Triangle Inequalities

4. On your graph paper, construct a right triangle with angle measures of 45°, 45°, and 90°.

   What is the measure of each exterior angle? _____

   What is the relationship between each exterior angle and the measure of each nonadjacent interior angle?

5. Based on your observations in problems 1–4, what is the relationship between each exterior angle of a triangle and its two nonadjacent interior angles?

Congruence
Set 6: Triangle Inequalities

Station 4

At this station, you will find graph paper, a protractor, a ruler, and six spaghetti noodles that are 1 inch, 2 inches, 2.5 inches, 3.5 inches, 4 inches, and 5 inches in length. Work as a group to construct the triangles and answer the questions.

1. On your graph paper, create two sides of a triangle with the 1-inch and 2-inch spaghetti noodles. Use the protractor to create a 30° angle between the two spaghetti noodles.

    Use your ruler and pencil to draw in the third side of the triangle. What is the length of this third side? _____

2. On the same graph, move the same 1-inch and 2-inch spaghetti noodles so that they have a 50° angle between them. Use your ruler and pencil to draw in the third side of this triangle.

    Is this third side longer than the third side you found in problem 1? _____

    What is the length of this third side? _____

3. In problems 1–2, you looked at two triangles with two congruent sides. What was the relationship between the included angle and the third side of each triangle?

4. On a new graph, use the 2.5-inch, 3.5-inch, and 5-inch spaghetti noodles to create a triangle. What is the measure of the angle opposite the 5-inch spaghetti noodle?

    _____

5. On a new graph, use the 2.5-inch, 3.5-inch, and 4-inch spaghetti noodles to create a triangle. What is the measure of the angle opposite the 4-inch spaghetti noodle?

    _____

    Is this angle less than or greater than the angle found in problem 4? _____

continued

Congruence
Set 6: Triangle Inequalities

6. In problems 4–5, you looked at two triangles with two congruent sides, but varying lengths of the third side. What was the relationship between the included angle and the third side of each triangle?

7. The relationship you found in problems 3 and 6 is called the **hinge theorem**. Why do you think the theorem has this name?

Congruence

Set 7: Translations, Dilations, Tessellations, and Symmetry

Instruction

Goal: To provide opportunities for students to develop concepts and skills related to reflections, rotations, dilations, and tessellations of polygons

Common Core Standards

Congruence

Experiment with transformations in the plane.

G-CO.1. Know precise definitions of angle, circle, perpendicular line, parallel line, and line segment, based on the undefined notions of point, line, distance along a line, and distance around a circular arc.

G-CO.3. Given a rectangle, parallelogram, trapezoid, or regular polygon, describe the rotations and reflections that carry it onto itself.

G-CO.4. Develop definitions of rotations, reflections, and translations in terms of angles, circles, perpendicular lines, parallel lines, and line segments.

G-CO.5. Given a geometric figure and a rotation, reflection, or translation, draw the transformed figure using, e.g., graph paper, tracing paper, or geometry software. Specify a sequence of transformations that will carry a given figure onto another.

Understand congruence in terms of rigid motions.

G-CO.6. Use geometric descriptions of rigid motions to transform figures and to predict the effect of a given rigid motion on a given figure; given two figures, use the definition of congruence in terms of rigid motions to decide if they are congruent.

Similarity, Right Triangles, and Trigonometry

Understand similarity in terms of similarity transformations.

G-SRT.1. Verify experimentally the properties of dilations given by a center and a scale factor:

    a. A dilation takes a line not passing through the center of the dilation to a parallel line, and leaves a line passing through the center unchanged.

    b. The dilation of a line segment is longer or shorter in the ratio given by the scale factor.

G-SRT.2. Given two figures, use the definition of similarity in terms of similarity transformations to decide if they are similar; explain using similarity transformations the meaning of similarity for triangles as the equality of all corresponding pairs of angles and the proportionality of all corresponding pairs of sides.

Congruence
Set 7: Translations, Dilations, Tessellations, and Symmetry

Instruction

Student Activities Overview and Answer Key

Station 1

Students will be given white computer paper, graph paper, scissors, tape, three toothpicks, and a ruler. Students will construct triangle, pentagon, and trapezoid "lollipops." They will perform reflections of these polygons. Then they will determine if the polygons are similar or congruent.

Answers

1. Answers will vary.
2. Answers will vary.
3. They are congruent because they are the same triangle; symmetric; the line of symmetry is the x-axis.
4. reflection
5. x-axis
6. Answers will vary; congruent; symmetric; the line of symmetry is the y-axis.
7. Answers will vary; congruent; symmetric; the line of symmetry is $y = x$.
8. Reflected figures are congruent. Reflected figures are symmetric.

Station 2

Students will be given white computer paper, graph paper, a cork board, scissors, push pins, a protractor, and a ruler. Students will work as a group to construct triangles, rectangles, and squares. They will perform rotations and determine congruence and rotational symmetry.

Answers

1. Answers will vary.
2. Answers will vary; congruent.
3. no, no, no, yes; 3; 120°, 240°, 360°; yes; 3
4. 2; 180°, 360°; yes; 2
5. yes; 4

Congruence
Set 7: Translations, Dilations, Tessellations, and Symmetry

Instruction

Station 3

Students will be given five toothpicks, five spaghetti noodles, graph paper, a ruler, and a protractor. Students will construct pentagons out of toothpicks and spaghetti noodles. They will use these pentagons to examine enlargement and reduction dilations. They will explain whether dilations yield congruent or similar figures.

Answers

1. Answers will vary. Sample answer: Each toothpick is 2 inches.
2. Answers will vary. Sample answer: Each piece of spaghetti is 4 inches.
3. Enlargement; answers will vary, but should be greater than 1. Sample answer: scale factor of 2
4. Reduction; answers will vary, but should be between 0 and 1. Sample answer: scale factor of 1/2
5. similar; Sample answer: The pentagons are not congruent because corresponding sides are not equal, but they are similar because corresponding sides are proportional.
6. scale factor greater than 1; scale factor between 0 and 1

Station 4

Students will be given a small square piece of poster board, a large piece of poster board, scissors, a ruler, tape, and colored markers. Students will create tessellations of an irregular figure and a hexagon. Students will determine if a tessellated figure is congruent or similar.

Answers

1. Answers will vary.
2. congruent
3.
4. Answers will vary.

Materials List/Setup

Station 1	white computer paper; graph paper; scissors; tape; three toothpicks; ruler
Station 2	white computer paper; graph paper; cork board; scissors; push pins; protractor; ruler
Station 3	five toothpicks; five spaghetti noodles; graph paper; ruler; protractor
Station 4	small square piece of poster board; large piece of poster board; scissors; ruler; tape; colored markers

Congruence
Set 7: Translations, Dilations, Tessellations, and Symmetry

Instruction

Discussion Guide

To support students in reflecting on the activities and to gather some formative information about student learning, use the following prompts to facilitate a class discussion to "debrief" the station activities.

Prompts/Questions

1. Are reflected polygons congruent or similar?
2. What is rotational symmetry?
3. Are rotated polygons congruent or similar?
4. Give an example of a polygon that has rotational symmetry of order 6.
5. What is dilation?
6. What is a scale factor?
7. What is a tessellation?

Think, Pair, Share

Have students jot down their own responses to questions, then discuss with a partner (who was not in their station group), and then discuss as a whole class.

Suggested Appropriate Responses

1. Reflected polygons are congruent.
2. A figure has rotational symmetry if it looks the same after a certain amount of rotation.
3. Rotated polygons are congruent.
4. A regular hexagon has rotational symmetry of order 6.
5. Dilation is enlargement or reduction of a figure by a scale factor.
6. The scale factor is the number which is multiplied by the size of the polygon to either enlarge or reduce its size.
7. A tessellation is a figure that covers the surface of a plane in a symmetrical manner with no overlaps or gaps.

Congruence
Set 7: Translations, Dilations, Tessellations, and Symmetry

Instruction

Possible Misunderstandings/Mistakes

- Not reflecting the polygon across the correct line of symmetry
- Not recognizing the order of a polygon in rotational symmetry
- Mixing up the possible values of the scale factors for enlargements and reductions
- Having gaps or overlaps in their tessellation

Congruence
Set 7: Translations, Dilations, Tessellations, and Symmetry

Station 1

At this station, you will find white computer paper, graph paper, scissors, tape, three toothpicks, and a ruler. Work as a group to construct the polygons and answer the questions.

On your graph paper, draw an *x*- and *y*-axis with the origin through the center of the paper.

On the white computer paper, draw an isosceles triangle with side lengths of 2 inches, 2 inches, and 1 inch. Cut the triangle out of the paper.

Tape the toothpick to the base of the triangle to create a triangle "lollipop."

1. Place the free end of the toothpick on the point (4, 0). Place the triangle so it is in the first quadrant.

    What are the vertices of the triangle?

2. Now place the free end of the toothpick on the point (4, 0) again, but place the triangle in the fourth quadrant.

    What are the vertices of this triangle?

3. Are the two triangles congruent or similar? Explain your answer.

    Are the two triangles symmetric? What is the line of symmetry?
    Explain your answer.

4. What transformation did you perform between the first and second triangle? Explain your answer.

continued

Congruence

Set 7: Translations, Dilations, Tessellations, and Symmetry

5. Which axis or point did you perform this transformation across? _____

6. Work as a group to construct a pentagon "lollipop." Reflect the pentagon across the y-axis.
   What are the vertices of the first pentagon?

   What are the vertices of the second pentagon?

   Are the two pentagons congruent or similar? Explain your answer.

   Are the two pentagons symmetric? What is the line of symmetry?
   Explain your answer.

7. Work as a group to construct a trapezoid. Reflect the trapezoid about the line $y = x$.
   What are the vertices of the first trapezoid?

   What are the vertices of the second trapezoid?

continued

Congruence
Set 7: Translations, Dilations, Tessellations, and Symmetry

Are the trapezoids congruent or similar? Explain your answer.

Are the trapezoids symmetric? What is the line of symmetry? Explain your answer.

8. Based on your observations in problems 1–7, do reflections create similar or congruent figures? Do reflections create symmetric figures? Explain your answer.

Congruence

Set 7: Translations, Dilations, Tessellations, and Symmetry

Station 2

At this station, you will find white computer paper, graph paper, a cork board, scissors, push pins, a protractor, and a ruler. Work as a group to construct the polygons and perform the transformations.

On your graph paper, draw an *x*- and *y*-axis with the origin through the center of the paper.

1. On the white computer paper, draw an equilateral triangle with side lengths of 1 inch. Cut the triangle out of the paper.

    Place one vertex of the triangle at the point (5, 5). Use the push pin and cork board to secure the vertex at this point.

    What are the vertices of the triangle?

2. Rotate the triangle 90° using the push pin as the point of rotation.

    What are the vertices of the triangle?

    Are the two triangles congruent or similar? Explain your answer.

    Remove the push pin.

3. Poke the push pin through the center of the triangle. Then push the pin into the coordinate (5, 5).

    Rotate the triangle 90° using the push pin as the point of rotation.

    Does the triangle look the same as it did before you rotated it? _____

    Rotate the triangle another 90° using the push pin as the point of rotation.

    Does the triangle look the same as it did before you rotated it? _____

continued

Congruence
Set 7: Translations, Dilations, Tessellations, and Symmetry

Rotate the triangle 90° using the push pin as the point of rotation.

Does the triangle look the same as it did before you rotated it? _____

Rotate the triangle another 90° using the push pin as the point of rotation.

Does the triangle look the same as it did before you rotated it? _____

How many times would the triangle match itself during a 360° rotation?

At what degrees of rotation would the triangle match itself?

Does the equilateral triangle have rotational symmetry? If so, what order? Explain your answer.

4. Repeat problem 3 for a rectangle.

How many times did the rectangle match itself during the 360° rotation?

At what degrees of rotation did the rectangle match itself?

Does the rectangle have rotational symmetry? If so, what order? Explain your answer.

5. Does a square have rotational symmetry? If so, what order? Explain your answer.

Congruence
Set 7: Translations, Dilations, Tessellations, and Symmetry

Station 3

At this station, you will find five toothpicks, five spaghetti noodles, graph paper, a ruler, and a protractor. Work as a group to construct the polygons and answer the questions.

Draw an *x*- and *y*-axis on your graph paper.

1. Use the toothpicks to create a pentagon with one of the vertices at the origin (0, 0).

   What is the length of each side of the toothpick pentagon? _____

2. Use the spaghetti noodles to create a pentagon with one of the vertices at the origin (0, 0).

   What is the length of each side of the spaghetti pentagon? _____

3. What type of dilation would you perform on the toothpick pentagon to create a pentagon the same size as the spaghetti pentagon? Explain your answer.

   What is the scale factor for this dilation? _____

4. What type of dilation would you perform on the spaghetti pentagon to create a pentagon the same size as the toothpick pentagon? Explain your answer.

   What is the scale factor for this dilation? _____

continued

Congruence
Set 7: Translations, Dilations, Tessellations, and Symmetry

5. Are the toothpick and spaghetti pentagons congruent or similar? Justify your answer.

6. Based on your observations in problems 1–5, what are the possible values of a scale factor in an enlargement?

   What are the possible values of a scale factor in a reduction?

NAME:

Congruence
Set 7: Translations, Dilations, Tessellations, and Symmetry

Station 4

At this station, you will find a small square piece of poster board, a large piece of poster board, scissors, a ruler, tape, and colored markers. Work as a group to create a tessellation.

On the square piece of poster board, draw a jagged line from one side to an opposite side.

Cut along the jagged line. Place the two pieces together so the straight edges of the square are lined together in the center. Tape the pieces together.

Trace the small figure you created onto the large poster board.

1. How can you line up the small figure so that it tessellates across the poster board so there are no overlaps or gaps?

Tessellate the small figure across the large poster board.

2. Are the tessellated figures congruent or similar? Explain your answer.

3. In the space below, show how you can tessellate a hexagon.

4. If there is time, color in your tessellation on the large poster board.

Congruence

Set 8: Coordinate Proof

Instruction

Goal: To provide opportunities for students to develop concepts and skills related to using coordinate geometry in order to prove that reflections, rotations, and translations yield congruent polygons and that dilations create similar polygons

Common Core Standards

Congruence

Experiment with transformations in the plane.

- **G-CO.3.** Given a rectangle, parallelogram, trapezoid, or regular polygon, describe the rotations and reflections that carry it onto itself.

- **G-CO.4.** Develop definitions of rotations, reflections, and translations in terms of angles, circles, perpendicular lines, parallel lines, and line segments.

- **G-CO.5.** Given a geometric figure and a rotation, reflection, or translation, draw the transformed figure using, e.g., graph paper, tracing paper, or geometry software. Specify a sequence of transformations that will carry a given figure onto another.

Understand congruence in terms of rigid motions.

- **G-CO.6.** Use geometric descriptions of rigid motions to transform figures and to predict the effect of a given rigid motion on a given figure; given two figures, use the definition of congruence in terms of rigid motions to decide if they are congruent.

Similarity, Right Triangles, and Trigonometry

Understand similarity in terms of similarity transformations.

- **G-SRT.1.** Verify experimentally the properties of dilations given by a center and a scale factor:

  a. A dilation takes a line not passing through the center of the dilation to a parallel line, and leaves a line passing through the center unchanged.

  b. The dilation of a line segment is longer or shorter in the ratio given by the scale factor.

Congruence
Set 8: Coordinate Proof

Instruction

Student Activities Overview and Answer Key

Station 1

Students will be given graph paper and a ruler. Students construct and reflect an irregular hexagon. They find the line of symmetry and determine if reflected figures are congruent.

Answers

1. (0, 3), (–1, 1), (–5, 1), (–6, 3), (–5, 5), (–1, 5); congruent
2. (0, –3), (1, –1), (5, –1), (6, –3), (5, –5), (1, –5); congruent
3. No; (0, –3), (–1, –1), (–5, –1), (–6, –3), (–5, –5), (–1, –5); congruent; $y = 1$

Station 2

Students will be given graph paper, a protractor, and a ruler. They will construct regular and irregular polygons. They will prove that rotational symmetry yields congruent figures. They will prove that the order of rotational symmetry for regular polygons equals the number of sides of the polygons.

Answers

1. Answers will vary. Sample answer:

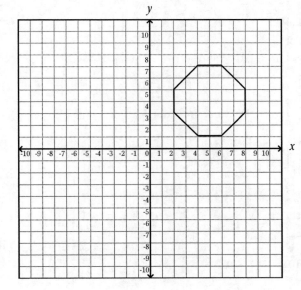

Congruence
Set 8: Coordinate Proof

Instruction

2. Answers will vary. Sample answer:

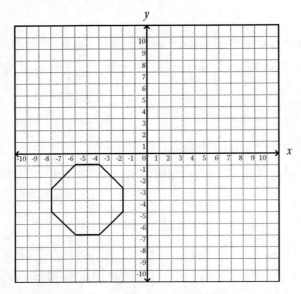

3. Yes, because you are only rotating the polygon.

4. Answers will vary. Sample graph:

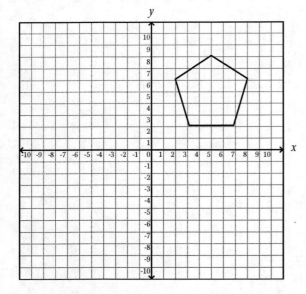

The pentagon matches in rotational symmetry 5 times.

Congruence
Set 8: Coordinate Proof

Instruction

5. Answers will vary. Sample graph:

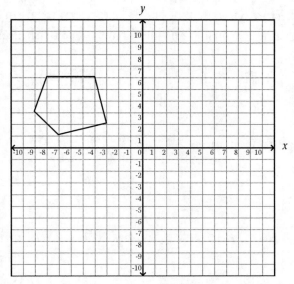

The pentagon matches in rotational symmetry 1 time.

6. For regular polygons, the number of matches during rotation is equal to the number of sides of the polygon. For irregular polygons, there is only one match.

Station 3

Students will be given graph paper and a ruler. They will construct a kite, then enlarge and reduce the kite. They will determine if dilations create similar or congruent figures.

Answers

1. kite
2. (4, 2), (8, 14), (0, 14), (4, 18); similar, corresponding sides are proportional
3. (4, 2), (5, 5), (3, 5), (4, 6); similar, corresponding sides are proportional
4. similar

Congruence
Set 8: Coordinate Proof

Instruction

Station 4

Students will be given graph paper and a ruler. They will construct and perform translations on irregular and regular polygons. They will determine if translations yield congruent or similar figures.

Answers

1. Scalene triangle; No, the sides aren't equal.
2. (1, 2), (5, 4), (1, 5); congruent
3. Yes, because all sides and angles are equal; congruent
4. (–3, –9), (1, –9), (–3, –5), (1, –5)
5. congruent

Materials List/Setup

Station 1   graph paper; ruler
Station 2   graph paper; protractor; ruler
Station 3   graph paper; ruler
Station 4   graph paper; ruler

Congruence
Set 8: Coordinate Proof

Instruction

Discussion Guide

To support students in reflecting on the activities and to gather some formative information about student learning, use the following prompts to facilitate a class discussion to "debrief" the station activities.

Prompts/Questions

1. How can you use coordinate geometry to prove that reflections yield congruent polygons?
2. How can you use coordinate geometry to prove rotational symmetry?
3. How can you use coordinate geometry to prove that dilations yield similar polygons?
4. How can you use coordinate geometry to prove that translations yield congruent polygons?

Think, Pair, Share

Have students jot down their own responses to questions, then discuss with a partner (who was not in their station group), and then discuss as a whole class.

Suggested Appropriate Responses

1. Reflect the polygon and verify that its size and shape remains the same.
2. Rotate the polygon and see if it matches itself more than once during a 360° rotation with the center of the polygon as the point of rotation.
3. Reduce or enlarge a polygon and verify that it is the same shape, but not the same size.
4. Translate a polygon and verify that its size and shape remain the same.

Possible Misunderstandings/Mistakes

- Not keeping the size and shape of the polygon the same when performing reflections
- Not recognizing how to find the "match" or order in rotational symmetry
- Not understanding the difference between regular and irregular polygons
- Not keeping the same shape when performing dilations
- Not keeping the same shape and size when performing translations

Congruence
Set 8: Coordinate Proof

Station 1

At this station, you will find graph paper and a ruler. Work as a group to construct the polygons and answer the questions.

Draw an *x*- and *y*-axis on your graph paper.

On your graph paper, construct an irregular hexagon that has vertices (1, 1), (5, 1), (6, 3), (5, 5), (1, 5), and (0, 3).

1. Reflect the figure across the *y*-axis.

   What are the vertices of this figure? _____

   Are the two figures congruent or similar? _____

2. Reflect the original figure across the *x*-axis.

   What are the vertices of this figure? _____

   Are the three figures congruent or similar? _____

3. Kathy says that the vertices of the figure from problem 1 reflected across the *x*-axis are (1, 1), (5, 1), (6, –3), (5, –5), (1, –5), and (0, –3). Is she correct? _____

   If not, what are the correct vertices? _____

   If Kathy didn't reflect the figure across the *x*-axis, what line of symmetry did she use? _____

   Perform this reflection on your graph paper to justify your answer.

Congruence
Set 8: Coordinate Proof

Station 2

At this station, you will find graph paper, a protractor, and a ruler. Work together to construct the polygons and answer the questions.

Draw an *x*- and *y*-axis on your graph paper.

1. On your graph paper, construct an octagon in the first quadrant.

   What are the vertices of the octagon?

   How many units on the graph paper can you shade in the octagon?

   _____

2. Rotate the octagon 180° about the origin.

   What are the vertices of this octagon?

   How many units on the graph paper can you shade in this octagon?

   _____

3. Should the number of units on the graph paper shaded in for each octagon be equal? Why or why not?

4. Construct a regular pentagon with side lengths of 5 units. (Each interior angle is 108°.)

   What are the vertices of this regular pentagon? _____

   Rotate the pentagon about its center.

   How many times does the pentagon match in rotational symmetry? _____

continued

Congruence
Set 8: Coordinate Proof

5. Construct an irregular pentagon in the second quadrant.

   What are the vertices of this irregular pentagon? _____

   Rotate the pentagon about its center.

   How many times does the pentagon match in rotational symmetry? _____

6. Based on your observations in problems 4 and 5, what can you say about rotational symmetry for regular polygons?

   What can you say about rotational symmetry for irregular polygons?

Congruence
Set 8: Coordinate Proof

Station 3

At this station, you will find graph paper and a ruler. Work as a group to construct the polygons and answer the questions.

Draw an *x*- and *y*-axis on your graph paper.

1. On your graph paper, construct a quadrilateral with vertices (4, 2), (6, 8), (4, 10), and (2, 8).

    What type of quadrilateral did you create? _____

2. Enlarge this quadrilateral by a scale factor of 2. Keep the vertex (4, 2) the same.

    What are the new vertices of the quadrilateral?

    Are the two quadrilaterals congruent or similar? Explain your answer.

3. Reduce the original quadrilateral from problem 1 by a scale factor of $\frac{1}{2}$. Keep the vertex (4, 2) the same.

    What are the new vertices of the quadrilateral? _____

    Are the three quadrilaterals congruent or similar? Explain your answer.

4. Based on your observations in problems 1–3, do dilations create congruent or similar figures?

    _____

Congruence
Set 8: Coordinate Proof

Station 4

At this station, you will find graph paper and a ruler. Work as a group to construct the polygons and answer the questions.

Draw an *x*- and *y*-axis on your graph paper.

1. On your graph paper, construct a triangle with vertices (–5, 2), (–1, 4), and (–5, 5).

    What type of triangle have you created? _____

    Is this triangle a regular polygon? Why or why not?

2. Translate the triangle 6 units to the right.

    What are the vertices of this triangle?

    Are the two triangles congruent or similar? _____

3. On a new graph, construct a square with vertices (1, 1), (5, 1), (1, 5), and (5, 5).

    Is this square a regular polygon? Why or why not?

4. Translate the square 10 units down and 4 units to the left.

    What are the vertices of this square?

    Are the two squares congruent or similar? _____

5. Based on your observations in problems 1–4, do translations yield congruent or similar figures? Justify your answer.

Congruence

Set 9: Rhombi, Squares, Kites, and Trapezoids

Instruction

Goal: To provide opportunities for students to develop concepts and skills related to describing and comparing relationships among quadrilaterals, including squares, rectangles, rhombi, parallelograms, trapezoids, and kites

Common Core Standards

Congruence

Experiment with transformations in the plane.

- **G-CO.1.** Know precise definitions of angle, circle, perpendicular line, parallel line, and line segment, based on the undefined notions of point, line, distance along a line, and distance around a circular arc.

Prove geometric theorems.

- **G-CO.11.** Prove theorems about parallelograms.

Make geometric constructions.

- **G-CO.12.** Make formal geometric constructions with a variety of tools and methods (compass and straightedge, string, reflective devices, paper folding, dynamic geometric software, etc.).

Student Activities Overview and Answer Key

Station 1

Students will be given four drinking straws, tape, a ruler, and a protractor. Students will construct a square. Then they will "lean" the square to create a rhombus. They describe why a rhombus is a parallelogram. Then they relate a square to a rhombus.

Answers

1. 90°
2. yes
3. yes
4. yes
5. yes
6. rhombus
7. A rhombus is like a square because all four sides are congruent.

Congruence
Set 9: Rhombi, Squares, Kites, and Trapezoids

Instruction

Station 2

Students will be given a ruler and a protractor. Students will construct a square. They will find the diagonals of the square. Then they will relate the square to a rectangle, a rhombus, and a parallelogram.

Answers

1.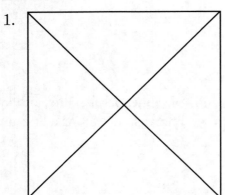

2. Pythagorean theorem; $2\sqrt{2}$

3. A square is always a rectangle. A square is a rectangle with four equal sides.

4. A square is a rhombus with four equal angles.

5. A square is a parallelogram with equal diagonals that bisect the angles.

Station 3

Students will be given three drinking straws, a ruler, tape, and scissors. Students will create a kite from the drinking straws. Students will describe the sides, angles, and diagonals in the kite. They will determine if a kite is a parallelogram.

Answers

1. Answers will vary.
2. Answers will vary.
3. Answers will vary.
4. Two pairs of adjacent sides are congruent.
5. Draw the horizontal diagonal to make the base of each isosceles triangle.
6. A kite is not a parallelogram because opposite sides are not congruent.
7. Diagonals are perpendicular.

Congruence
Set 9: Rhombi, Squares, Kites, and Trapezoids

Instruction

Station 4

Students will be given tracing paper, a ruler, and scissors. Students will construct and describe a trapezoid. They will explain if the trapezoid can be turned into a parallelogram. They will construct and describe an isosceles trapezoid. They will explain if the trapezoid can be turned into a parallelogram.

Answers

1. Yes, because it has four sides.
2. No, because opposite sides are not congruent.
3. No, because the sides of the trapezoid consist of two noncongruent triangles and a rectangle.
4. Yes, because it has four sides.
5. No, because opposite sides are not congruent.
6. Yes, because the trapezoid consists of two congruent isosceles triangles and a rectangle; rectangle

Materials List/Setup

Station 1	four drinking straws; tape; ruler; protractor
Station 2	ruler; protractor
Station 3	three drinking straws; ruler; tape; scissors
Station 4	tracing paper; ruler; scissors

Congruence
Set 9: Rhombi, Squares, Kites, and Trapezoids

Instruction

Discussion Guide

To support students in reflecting on the activities and to gather some formative information about student learning, use the following prompts to facilitate a class discussion to "debrief" the station activities.

Prompts/Questions

1. What is a rhombus? Is it a parallelogram?
2. How does a square relate to a rhombus?
3. What is a kite? Is it a parallelogram?
4. What is a trapezoid? Is it a parallelogram?

Think, Pair, Share

Have students jot down their own responses to questions, then discuss with a partner (who was not in their station group), and then discuss as a whole class.

Suggested Appropriate Responses

1. A rhombus is a quadrilateral with all four sides equal and opposite angles equal. Yes, it is a parallelogram.
2. A square is a rhombus with four right angles.
3. A kite is a quadrilateral that has two pairs of adjacent sides that are equal. No, it is not a parallelogram.
4. A trapezoid is a quadrilateral that has one pair of parallel sides. No, it is not a parallelogram.

Possible Misunderstandings/Mistakes

- Not recognizing that a square is a type of rhombus
- Not recognizing that all sides are equal in a rhombus
- Not realizing that all squares are rectangles, but not all rectangles are squares
- Not recognizing that pairs of adjacent sides in a kite have equal length

NAME:

Congruence
Set 9: Rhombi, Squares, Kites, and Trapezoids

Station 1

At this station, you will find four drinking straws, tape, a ruler, and a protractor. Work together to answer the questions.

1. Tape the four drinking straws together to create a square.

   What are the measures of each interior angle? _____

2. "Lean" the straws so that you create interior angles of 120°, 60°, 120°, and 60°.

   Are the lengths of the sides still equal? Why or why not?

3. Are pairs of opposite sides congruent? _____

4. Are pairs of opposite angles congruent? _____

5. Is this figure a parallelogram? Why or why not?

6. What is the special name for this quadrilateral? _____

7. How does this quadrilateral relate to a square?

Congruence

Set 9: Rhombi, Squares, Kites, and Trapezoids

Station 2

At this station, you will find a ruler and a protractor. Work together to construct the figures and answer the questions.

1. In the space below, construct a square with side lengths of 2 inches. Construct the diagonals on the square.

2. How can you find the length of each diagonal without using your ruler?

   What is the length of each diagonal? _____

   Verify your answer with your ruler.

3. Is a square always a rectangle or is a rectangle always a square? Justify your answer.

4. How can you relate a square to a rhombus?

continued

Congruence
Set 9: Rhombi, Squares, Kites, and Trapezoids

In the space below, draw a square and rhombus to show how they relate to each other.

5. How can you relate a square to a parallelogram?

Congruence

Set 9: Rhombi, Squares, Kites, and Trapezoids

Station 3

At this station, you will find three drinking straws, a ruler, tape, and scissors. Work as a group to create a kite.

1. Cut one straw exactly in half. Tape the four straws together to create a kite similar to the image below.

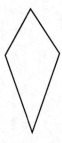

2. What is the length of each side? _____

3. What is the measure of each angle? _____

4. What is the relationship between adjacent sides of a kite?

5. How can you create two isosceles triangles from the kite?

6. Is a kite a parallelogram? Why or why not?

continued

Congruence

Set 9: Rhombi, Squares, Kites, and Trapezoids

7. Place your straw kite in the space below. Draw the diagonals of the kite. What relationship do the diagonals have to each other?

Congruence

Set 9: Rhombi, Squares, Kites, and Trapezoids

Station 4

At this station, you will find tracing paper, a ruler, and scissors. Work as a group to answer the questions.

Given the trapezoid:

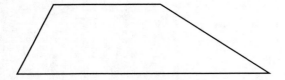

Use the tracing paper to trace the trapezoid above. Cut the trapezoid out of the tracing paper.

1. Is this trapezoid a quadrilateral? Why or why not?

2. Is this trapezoid a parallelogram? Why or why not?

   If so, what type of parallelogram did you create? _____

3. Can you cut the trapezoid into two pieces and create a parallelogram? Why or why not?

continued

Congruence
Set 9: Rhombi, Squares, Kites, and Trapezoids

Given the trapezoid:

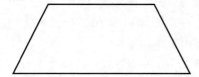

Use the tracing paper to trace the trapezoid above. Cut the trapezoid out of the tracing paper.

4. Is this trapezoid a quadrilateral? Why or why not?

5. Is this trapezoid a parallelogram? Why or why not?

6. Can you cut the trapezoid into two pieces and create a parallelogram? Why or why not?

   If so, what type of parallelogram did you create? _____

Similarity, Right Triangles, and Trigonometry

Set 1: Similarity and Scale Factor

Instruction

Goal: To provide opportunities for students to develop concepts and skills related to proving that triangles are similar using scale factors and angle relationships

Common Core Standards

Congruence

Experiment with transformations in the plane.

G-CO.1. Know precise definitions of angle, circle, perpendicular line, parallel line, and line segment, based on the undefined notions of point, line, distance along a line, and distance around a circular arc.

Similarity, Right Triangles, and Trigonometry

Understand similarity in terms of similarity transformations.

G-SRT.1. Verify experimentally the properties of dilations given by a center and a scale factor:

    a. A dilation takes a line not passing through the center of the dilation to a parallel line, and leaves a line passing through the center unchanged.

    b. The dilation of a line segment is longer or shorter in the ratio given by the scale factor.

G-SRT.2. Given two figures, use the definition of similarity in terms of similarity transformations to decide if they are similar; explain using similarity transformations the meaning of similarity for triangles as the equality of all corresponding pairs of angles and the proportionality of all corresponding pairs of sides.

Prove theorems involving similarity.

G-SRT.5. Use congruence and similarity criteria for triangles to solve problems and to prove relationships in geometric figures.

Expressing Geometric Properties with Equations

Use coordinates to prove simple geometric theorems algebraically.

G-GPE.4. Use coordinates to prove simple geometric theorems algebraically.

G-GPE.7. Use coordinates to compute perimeters of polygons and areas of triangles and rectangles, e.g., using the distance formula.★

Similarity, Right Triangles, and Trigonometry
Set 1: Similarity and Scale Factor

Instruction

Student Activities Overview and Answer Key

Station 1

Students will be given graph paper, a ruler, a red marker, a green marker, and a blue marker. Students will construct similar and non-similar rectangles. They will determine ratios of corresponding sides to see if the rectangles are similar. Then they will find the relationship between perimeter and area with the corresponding sides of similar rectangles.

Answers

1. red rectangle:

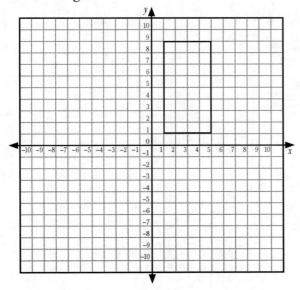

2. green rectangle:

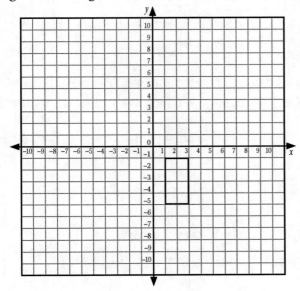

Similarity, Right Triangles, and Trigonometry
Set 1: Similarity and Scale Factor

Instruction

3. blue rectangle:

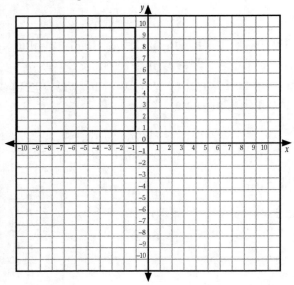

4. 4/2; 8/4; 4/2 = 8/4

5. 4/9; 8/10; 4/9 ≠ 8/10

6. red and green

7. red = 24 units, green = 12 units, blue = 38 units

8. red = 32 square units, green = 8 square units, blue = 90 square units

9. The perimeter has the same ratio as the corresponding sides.

10. The area is equal to the square of the ratio of the corresponding sides.

Station 2

Students will be given a ruler and a protractor. They will measure the angles and lengths of two trapezoids. They will determine if the two trapezoids are similar. Students will derive the relationship between the angles in similar polygons.

Answers

1. angles B and A = 56°, angles C and D = 124°

2. angles P and O = 56°, angles M and N = 124°

3. no

4. The base = 3 inches, the sides = 1.3 inches each, and the top = 1.7 inches.

Similarity, Right Triangles, and Trigonometry
Set 1: Similarity and Scale Factor

Instruction

5. The base = 1.5 inches, the sides = 0.65 inches each, and the top = 0.85 inches.
6. 2/1
7. Yes. 3/1.5 = 1.7/0.85 = 1.3/0.65
8. Answers will vary.
9. Answers will vary.
10. Corresponding angles are equal.

Station 3

Students will be given graph paper, a blue marker, a red marker, and a green marker. They will create similar rectangles using graph paper. Students will find the scale factors of the rectangles by physically placing them on top of each other. They will find the scale factors of other rectangles.

Answers

1. Student drawings should depict a blue rectangle that is 12 units long by 9 units wide.
2. Student drawings should depict a red rectangle that is 4 units long by 3 units wide.
3. 3; 3; 1/3; 1/3
4. 3; 1/3
5. 60 units long, 45 units wide
6. 2 units long, 1.5 units wide
7. No, because the rectangles aren't similar. No, because the rectangles aren't similar.
8. Answers will vary. Possible answer: Scale drawings used in architecture.

Station 4

Students will be given a ruler and a protractor. They will find the missing side lengths of similar triangles. Students will discuss the strategy they used to determine whether triangles were similar or congruent. Students will explain why right triangles are not similar to obtuse triangles. Then they will explain why not all right triangles are similar.

Answers

1. $x = 16$, similar
2. $x = 6$, similar

Similarity, Right Triangles, and Trigonometry
Set 1: Similarity and Scale Factor

Instruction

3. $x = 3$, congruent
4. Answers will vary.
5. Answers will vary.
6. no
7. no

Materials List/Setup

Station 1   graph paper; ruler; red marker; green marker; blue marker
Station 2   ruler; protractor
Station 3   graph paper; scissors; blue marker; red marker; green marker
Station 4   ruler; protractor

Similarity, Right Triangles, and Trigonometry
Set 1: Similarity and Scale Factor

Instruction

Discussion Guide

To support students in reflecting on the activities and to gather some formative information about student learning, use the following prompts to facilitate a class discussion to "debrief" the station activities.

Prompts/Questions
1. What is the relationship between corresponding sides in similar polygons?
2. What is the relationship between corresponding angles in similar polygons?
3. What does it mean to scale up a figure by a factor of 4?
4. What does it mean to scale down a figure by a factor of 1/2?
5. How can you find the missing side length in similar triangles?

Think, Pair, Share

Have students jot down their own responses to questions, then discuss with a partner (who was not in their station group), and then discuss as a whole class.

Suggested Appropriate Responses
1. All corresponding sides are proportional.
2. All corresponding angles are congruent.
3. Make the figure 4 times the size of the original.
4. Make the figure 1/2 the size of the original.
5. Use proportions of corresponding sides.

Possible Misunderstandings/Mistakes
- Not correctly identifying corresponding sides of similar polygons
- Not setting up proportions correctly between similar polygons
- Not multiplying scale factor by all sides in the polygon
- Not realizing that not all right triangles are similar because they can have different corresponding angles

Similarity, Right Triangles, and Trigonometry
Set 1: Similarity and Scale Factor

Station 1

At this station, you will find graph paper, a ruler, a red marker, a green marker, and a blue marker. Work as a group to construct the polygons and answer the questions.

1. On your graph paper, use the red marker to construct a rectangle that has vertices (1, 1), (5, 1), (1, 9), and (5, 9).

2. On your graph paper, use the green marker to construct a rectangle that has vertices (1, −1), (3, −1), (1, −5), and (3, −5).

3. On your graph paper, use the blue marker to construct a rectangle that has vertices (−1, 1), (−1, 10), (−11, 10), and (−11, 1).

4. What is the ratio between the shorter side of the red rectangle and the shorter side of the green rectangle?

    _____

    What is the ratio between the longer side of the red rectangle and the longer side of the green rectangle?

    _____

    What is the proportion between the shorter side and the longer side of the red and green triangles?

    _____

5. What is the ratio between the shorter side of the red rectangle and shorter side of the blue rectangle?

    _____

    What is the ratio between the longer side of the red rectangle and the longer side of the blue rectangle?

    _____

    What is the proportion between the shorter side and the longer side of the red and blue rectangles?

    _____

continued

Similarity, Right Triangles, and Trigonometry
Set 1: Similarity and Scale Factor

6. Which of the rectangles are similar? Write your answer as a proportion. Justify your answer by measuring and comparing the side lengths of each rectangle. Show your work and answer in the space below.

7. What is the perimeter of each rectangle? Show your work and answer in the space below.

8. What is the area of each rectangle? Show your work and answer in the space below.

9. Based on your observations in problems 1–8, what is the relationship between the perimeter and corresponding sides of similar rectangles?

10. Based on your observations in problems 1–8, what is the relationship between the area and the corresponding sides of similar rectangles?

Similarity, Right Triangles, and Trigonometry
Set 1: Similarity and Scale Factor

Station 2

At this station, you will find a ruler and a protractor. Work as a group to answer the questions.

For problems 1–2, find the measure of each angle.

1.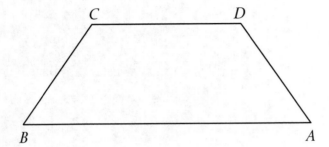

   ∠A = _____   ∠B = _____   ∠C = _____   ∠D = _____

2.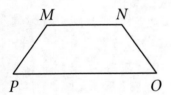

   ∠M = _____   ∠N = _____   ∠O = _____   ∠P = _____

3. Can you say that the trapezoids in problems 1 and 2 are similar based solely on the measure of the angles? Why or why not?

4. What is the length of each side in problem 1? _____

5. What is the length of each side in problem 2? _____

continued

Similarity, Right Triangles, and Trigonometry
Set 1: Similarity and Scale Factor

6. What is the ratio of the corresponding sides between problems 1 and 2?
   _____

7. Are the trapezoids in problems 1 and 2 similar? Explain your answer by setting up a proportion.

8. In the space below, construct and label a trapezoid that is similar to the trapezoid in problem 1.

9. In the space below, construct and label a trapezoid that is NOT similar to the trapezoid in problem 1.

10. What is the relationship between the angles in similar trapezoids and other similar polygons?
    _____

Similarity, Right Triangles, and Trigonometry
Set 1: Similarity and Scale Factor

Station 3

At this station, you will find graph paper, scissors, a blue marker, a red marker, and a green marker. Work as a group to answer the questions.

1. On your graph paper, construct a rectangle that is 12 units long by 9 units wide. Shade in the rectangle with your blue marker.

2. On your graph paper, construct a rectangle that is 4 units long by 3 units wide. Shade in the rectangle with your red marker.

3. Use your scissors to cut out each rectangle from the graph paper. Place the red rectangle on the blue rectangle.

   How many red rectangles can you fit along the length of the blue rectangle?

   _____

   How many red rectangles can you fit along the width of the blue rectangle?

   _____

   Place the blue rectangle on the red rectangle.

   What fraction of the blue rectangle can you fit along the length of the red rectangle?

   _____

   What fraction of the blue rectangle can you fit along the width of the red rectangle?

   _____

4. In problem 3, you found the scale factors of the two rectangles.

   The blue rectangle was larger than the red rectangle by what scale factor?

   _____

   The red rectangle was smaller than the blue rectangle by what scale factor?

   _____

continued

Similarity, Right Triangles, and Trigonometry
Set 1: Similarity and Scale Factor

5. What would be the length and width of a rectangle that is the blue rectangle scaled up by a factor of 5? _____

6. What would be the length and width of a rectangle that is the red rectangle scaled down by a factor of $\frac{1}{2}$? _____

7. On your graph paper, construct a rectangle that is 10 units long by 2 units wide. Shade in the rectangle with your green marker.

    Use your scissors to cut this rectangle from the graph paper.

    Can you find the scale factor between the green rectangle and the blue rectangle? Why or why not?

    Can you find the scale factor between the green rectangle and the red rectangle? Why or why not?

8. What is an example of a real-world application of scale factors?

Similarity, Right Triangles, and Trigonometry
Set 1: Similarity and Scale Factor

Station 4

At this station, you will find a ruler and a protractor. Work as a group to answer the questions.

For problems 1–3, the triangles are similar. Find the missing length, x. Then use your ruler to measure the sides and check the ratios of corresponding sides to verify that the triangles are similar. (Drawings are not to scale.)

1.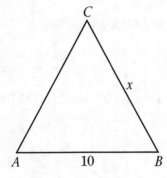

2.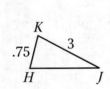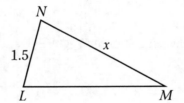

3.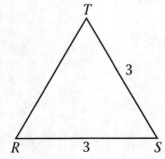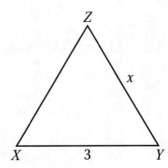

continued

124

Similarity, Right Triangles, and Trigonometry
Set 1: Similarity and Scale Factor

4. What strategy did you use to determine the missing side length, *x* in each of the pairs of triangles on the previous page?

5. What strategy did you use to determine whether the triangles were congruent or similar?

6. Is a right triangle similar to an obtuse triangle? Use your ruler and protractor to help explain your answer in the space below.

7. Is a right triangle similar to all right triangles? Use your ruler and protractor to help explain your answer in the space below.

Similarity, Right Triangles, and Trigonometry

Set 2: Ratio Segments

Instruction

Goal: To provide opportunities for students to develop concepts and skills related to theorems involving segments divided proportionally in triangles and transversals through parallel lines

Common Core Standards

Congruence

Experiment with transformations in the plane.

G-CO.1. Know precise definitions of angle, circle, perpendicular line, parallel line, and line segment, based on the undefined notions of point, line, distance along a line, and distance around a circular arc.

Prove geometric theorems.

G-CO.10. Prove theorems about triangles.

Make geometric constructions.

G-CO.12. Make formal geometric constructions with a variety of tools and methods (compass and straightedge, string, reflective devices, paper folding, dynamic geometric software, etc.).

Similarity, Right Triangles, and Trigonometry

Prove theorems involving similarity.

G-SRT.4. Prove theorems about triangles.

G-SRT.5. Use congruence and similarity criteria for triangles to solve problems and to prove relationships in geometric figures.

Expressing Geometric Properties with Equations

Use coordinates to prove simple geometric theorems algebraically.

G-GPE.4. Use coordinates to prove simple geometric theorems algebraically.

Student Activities Overview and Answer Key

Station 1

Students will be given a ruler and a protractor. Students will construct an equilateral triangle and a line parallel to one side of the triangle. They will derive a relationship between the two triangles. They will repeat this process for a right triangle. Students will find that a line inside the triangle that is parallel to one side of the triangle will create two similar triangles.

Similarity, Right Triangles, and Trigonometry
Set 2: Ratio Segments

Instruction

Answers

1.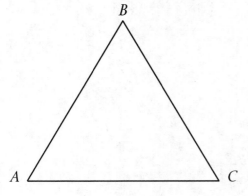

2.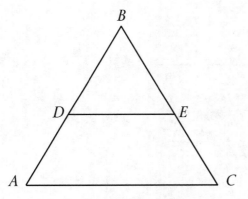

3. Answers will vary.

4. Triangles are similar.

5.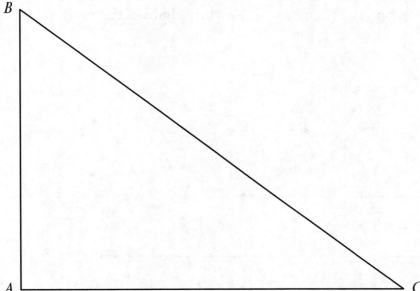

Similarity, Right Triangles, and Trigonometry
Set 2: Ratio Segments

Instruction

6.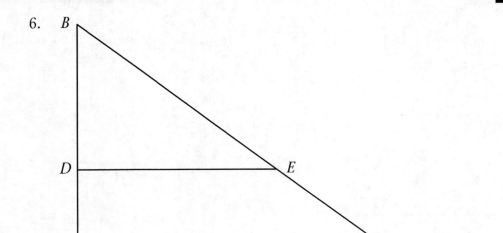

7. Answers will vary.

8. Triangles are similar.

9. You create two triangles that are similar.

Station 2

Students will be given a ruler. Students will construct a triangle and a line parallel to one of the sides of the triangle. They will realize that if a line is parallel to one side of a triangle then it divides the other two sides proportionately. Then they will find the lengths of the missing sides using this principle.

Answers

1. Answers will vary. Possible answer:

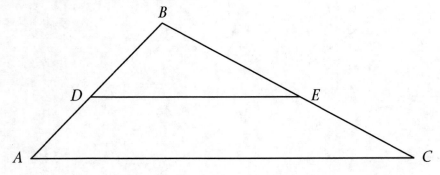

Similarity, Right Triangles, and Trigonometry
Set 2: Ratio Segments

Instruction

   2. Answers will vary.
   3. Corresponding sides are proportional.
   4. Answers will vary, but corresponding sides should be in proportion to one another.
   5. $DB = 15$
   6. $x = 12$; $EC = 18$ and $BE = 12$

Station 3

Students will be given a ruler, a compass, and a protractor. Students will construct angle bisectors for an obtuse triangle. They will determine the relationship between the segments opposite the angle bisector and the sides that form the bisected angle. Then they will find missing side lengths based on this principle.

Answers

1.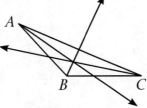

   Answers will vary.

2. If a ray bisects one angle of a triangle, then it divides the opposite side into segments whose lengths are proportional to the lengths of the two sides that form the bisected angle.

3. $x = 36$

4. $x = 3$; sides are 10 and 6

Station 4

Students will be given graph paper and a ruler. Students will construct three parallel lines cut by a pair of transversals. They will measure the segments of the transversals cut by the parallel lines. They will realize that when three (or more) parallel lines are cut by a pair of transversals, the transversals are divided proportionally by the parallel lines.

Similarity, Right Triangles, and Trigonometry
Set 2: Ratio Segments

Instruction

Answers

1.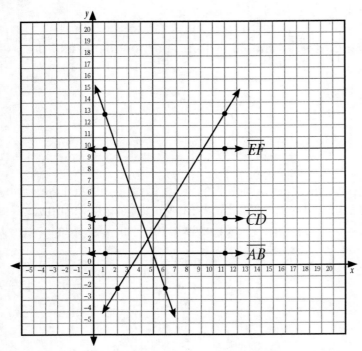

2. Answers may vary. Sample answers: 3.16; 6.32

3. Answers may vary. Sample answers: 3.61; 7.21

4. Answers may vary but a proportion should exist. Sample answers: 3.16/6.32 = 3.61/7.21 = 1/2

5. Yes; the third transversal is divided proportionally to the other two transversals.

6. The transversals are divided proportionally by the parallel lines.

Materials List/Setup

Station 1   ruler; protractor

Station 2   ruler

Station 3   ruler; compass; protractor

Station 4   graph paper; ruler

Similarity, Right Triangles, and Trigonometry
Set 2: Ratio Segments

Instruction

Discussion Guide

To support students in reflecting on the activities and to gather some formative information about student learning, use the following prompts to facilitate a class discussion to "debrief" the station activities.

Prompts/Questions

1. What is the relationship between two triangles created by a line inside the triangle that is parallel to one of the sides of the triangle?

2. What is the relationship between the sides of the two triangles created by a line inside the triangle that is parallel to one of the sides of the triangle?

3. What is the relationship between an angle bisector of a triangle and the opposite side and the lengths of the two sides that bisect the angle?

4. What is the relationship between three or more parallel lines cut by a pair of transversals?

Think, Pair, Share

Have students jot down their own responses to questions, then discuss with a partner (who was not in their station group), and then discuss as a whole class.

Suggested Appropriate Responses

1. The triangles are similar.

2. The sides are proportional.

3. The opposite side is divided into segments whose lengths are proportional to the lengths of the two sides that form the bisected angle.

4. When three (or more) parallel lines are cut by a pair of transversals, the transversals are divided proportionally by the parallel lines.

Possible Misunderstandings/Mistakes

- Incorrectly setting up proportions to determine side lengths of similar triangles
- Incorrectly setting up proportions to determine lengths of transversal segments through three or more parallel lines
- Incorrectly bisecting an angle

NAME: _____

Similarity, Right Triangles, and Trigonometry
Set 2: Ratio Segments

Station 1

At this station, you will find a ruler and a protractor. Work as a group to answer the questions.

1. In the space below, construct an equilateral triangle with side lengths of 2 inches. Label the vertices of the triangle as A, B, and C.

2. Construct a horizontal line segment inside the triangle that is parallel to base $\overline{AC}$ of the triangle. Label the endpoints of the line segment as D and E, with D on $\overline{AB}$ and E on $\overline{BC}$.

3. Find the following measurements.

    DB = _____

    BE = _____

4. What is the relationship between △ABC and △DBE?

    _____

5. In the space below, construct a right triangle with side lengths of 3 inches, 4 inches, and 5 inches. Label the vertices of the triangle as A, B, and C.

6. Construct a horizontal line segment inside the triangle that is parallel to side $\overline{AC}$ of the triangle. Label the endpoints of the line segment as D and E, with D on $\overline{AB}$ and E on $\overline{BC}$.

continued

Similarity, Right Triangles, and Trigonometry
Set 2: Ratio Segments

7. Find the following measurements.

   DB = _____

   BE = _____

8. What is the relationship between △ABC and △DBE?

   _____

9. Based on your observations in problems 1–8, what is created when you cut a triangle by a line parallel to a side of the triangle?

   _____

Similarity, Right Triangles, and Trigonometry
Set 2: Ratio Segments

Station 2

At this station, you will find a ruler. Work as a group to construct the triangles and answer the questions.

1. In the space below, draw a triangle with side lengths 2 inches, 3 inches, and 4 inches. Label the triangle *ABC*.

   Draw a line parallel to *AC* inside the triangle that intersects the two sides *AB* and *BC*. Label the end points of the line as *D* and *E*, with *D* on *AB* and *E* on *BC*.

2. Find the following measurements:

   AD = _____

   DB = _____

   BE = _____

   EC = _____

3. What is the relationship between corresponding sides of △*ABC*:△*DBE*?

4. What proportion represents the relationship between corresponding sides of △*ABC*:△*BDE*?

   _____

continued

Similarity, Right Triangles, and Trigonometry
Set 2: Ratio Segments

5. In the triangle below, $AD = 5$, $EC = 8$, and $BE = 24$. What is the length of DB? Show your work and answer in the space below.

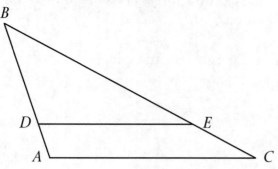

6. In the triangle below, $AB = 10$, $DB = 4$, $EC = x + 6$, and $BE = x$. What are the lengths of EC and BE? Show your work and answer in the space below.

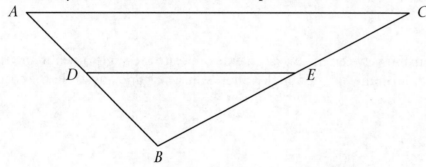

Similarity, Right Triangles, and Trigonometry
Set 2: Ratio Segments

Station 3

At this station, you will find a ruler, a compass, and a protractor. Work as a group to construct the triangles and angle bisectors, and answer the questions.

1. In the space below, construct an obtuse triangle. Label the triangle *ABC*.

   Construct the angle bisectors of the triangle.

   What are the lengths of the segments opposite each angle bisector? Write these lengths on your triangle.

2. What is the relationship between the two segments opposite the angle bisector and the length of the two sides that form the bisected angle? Show your work and answer in the space below.

3. The illustration below shows the angle bisector of $\angle B$. What is the value of *x*? Show your work and answer in the space below.

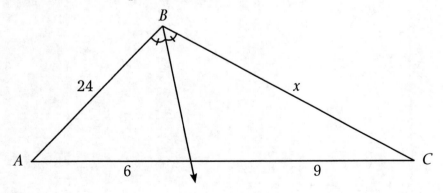

continued

Similarity, Right Triangles, and Trigonometry
Set 2: Ratio Segments

4. The illustration below shows the angle bisector of $\angle B$. What is the value of x? What is the value of each side? Show your work and answer in the space below.

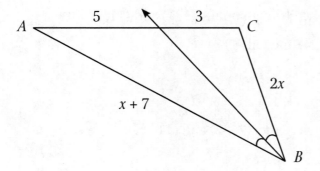

Similarity, Right Triangles, and Trigonometry
Set 2: Ratio Segments

Station 4

At this station, you will find graph paper and a ruler. Work as a group to answer the questions.

1. On your graph paper, construct line *AB* through points (1, 1) and (11, 1).

    Construct line *CD* through points (1, 4) and (11, 4).

    Construct line *EF* through points (1, 10) and (11, 10).

    Construct a transversal through points (1, 13) and (6, –2).

    Construct a second transversal through points (2, –2) and (11, 13).

2. What is the length of the first transversal between $\overline{AB}$ and $\overline{CD}$? _____

    What is the length of the first transversal between $\overline{CD}$ and $\overline{EF}$? _____

3. What is the length of the second transversal between $\overline{AB}$ and $\overline{CD}$? _____

    What is the length of the second transversal between $\overline{CD}$ and $\overline{EF}$? _____

4. What is the relationship between the segments of each transversal? Explain your answer.

5. Construct another line parallel to $\overline{AB}$. Does the relationship you created in problem 4 still apply? Explain your answer.

6. In general, when three or more parallel lines are cut by a pair of transversals, what effect(s) do the parallel lines have on the transversals?

Similarity, Right Triangles, and Trigonometry

Set 3: Sine, Cosine, and Tangent Ratios, and Angles of Elevation and Depression

Instruction

Goal: To provide opportunities for students to develop concepts and skills related to trigonometric ratios for right triangles and angles of elevation and depression

Common Core Standards

Congruence

Experiment with transformations in the plane.

G-CO.1. Know precise definitions of angle, circle, perpendicular line, parallel line, and line segment, based on the undefined notions of point, line, distance along a line, and distance around a circular arc.

Similarity, Right Triangles, and Trigonometry

Define trigonometric ratios and solve problems involving right triangles.

G-SRT.6. Understand that by similarity, side ratios in right triangles are properties of the angles in the triangle, leading to definitions of trigonometric ratios for acute angles.

G-SRT.7. Explain and use the relationship between the sine and cosine of complementary angles.

G-SRT.8. Use trigonometric ratios and the Pythagorean theorem to solve right triangles in applied problems.★

Student Activities Overview and Answer Key

Station 1

Students will be given a graphing calculator, a ruler, and a protractor. Students will find the sine, cosine, and tangent ratios in a right triangle. Then they will find the measure of each angle in the triangle using these trigonometric ratios.

Answers

1. $\dfrac{3}{5}$
2. $\dfrac{4}{5}$
3. $\dfrac{3}{4}$
4. $\dfrac{4}{5}$

Similarity, Right Triangles, and Trigonometry

Set 3: Sine, Cosine, and Tangent Ratios, and Angles of Elevation and Depression

Instruction

5. $\dfrac{3}{5}$

6. $\dfrac{4}{3}$

7. $\alpha = 36.9°; \beta = 53.1°$

Station 2

Students will be given a graphing calculator. Students will work together to find the length of a missing side given the length of one side and a reference angle. Then they will use the trigonometric ratios to find the length of the missing side.

Answers

1. $\sin 50° = \dfrac{x}{12}$
2. $\sin 50° = x/12; \ 0.766 = x/12; \ x = 9.19$
3. $\sin 45° = \dfrac{1}{\sqrt{2}}$
4. $x = 12.85$
5. $x = 20.31$
6. $x = 2.13$

Station 3

Students will be given three drinking straws, a graphing calculator, scissors, tape, and a ruler. Students will construct a right triangle out of the drinking straws. They will use it to model an angle of elevation problem. They will use trigonometric ratios of right triangles to find the angle of elevation.

Answers

1. Answers will vary. Sample answer: 5 and 7 inches
2. Answers will vary. Sample answer: approximately 8.6 inches
3. Answers will vary; leg
4. Answers will vary. Sample answer: 8.6 inches; hypotenuse
5. Answers will vary. Sample answer: 7 inches; leg
6. Answers will vary. Possible method: $\tan(x) = 5/7; \ x = 35.5$

Similarity, Right Triangles, and Trigonometry
Set 3: Sine, Cosine, and Tangent Ratios, and Angles of Elevation and Depression

Instruction

Station 4

Students will be given a tape measure and tape. Students physically model an angle of depression. They use trigonometric ratios of right triangles to find the angle of depression.

Answers

1. Answers will vary. Sample answer:

   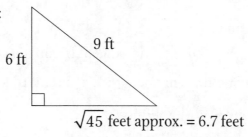

2. Answers will vary. Sample answer: $\cos(x) = 6/9$; $\cos^{-1} = \cos^{-1}(6/9)$; x approx. = 48.2

3. Answers will vary; Sample answer: It is the angle that is "looking downward."

4. Answers will vary. Sample answer:

   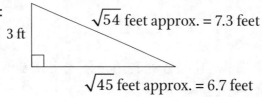

5. Answers will vary. Sample answer: $\cos(x) = 3/7$; $\cos^{-1} = \cos^{-1}(3/7)$; x approx. = 64.6

6. The angle of depression found in problem 2 is less than the one found in problem 5.

Materials List/Setup

Station 1	graphing calculator; ruler; protractor
Station 2	graphing calculator
Station 3	three drinking straws; graphing calculator; scissors; tape; ruler
Station 4	tape measure; tape

Similarity, Right Triangles, and Trigonometry
Set 3: Sine, Cosine, and Tangent Ratios, and Angles of Elevation and Depression

Instruction

Discussion Guide

To support students in reflecting on the activities and to gather some formative information about student learning, use the following prompts to facilitate a class discussion to "debrief" the station activities.

Prompts/Questions

1. What are the sine, cosine, and tangent trigonometric ratios for right triangles?
2. How do you find the length of a side of a right triangle given one side and a reference angle?
3. What is the angle of elevation?
4. How do you find the angle of elevation?
5. What is the angle of depression?
6. How do you find the angle of depression?

Think, Pair, Share

Have students jot down their own responses to questions, then discuss with a partner (who was not in their station group), and then discuss as a whole class.

Suggested Appropriate Responses

1. $\sin\theta = \dfrac{\text{opposite}}{\text{hypotenuse}}$; $\cos\theta = \dfrac{\text{adjacent}}{\text{hypotenuse}}$; $\tan\theta = \dfrac{\text{opposite}}{\text{adjacent}}$
2. Use the appropriate trigonometric ratio to write an equation and solve the equation for the missing side.
3. The angle of elevation is the angle between the horizontal and the line of sight of an object above the horizontal.
4. Use the information to draw a right triangle. Then use the appropriate trigonometric ratio to write an equation. Solve the equation for the missing value.
5. The angle of depression is the angle between the horizontal and the line of sight of an object below the horizontal.
6. Use the information to draw a right triangle. Then use the appropriate trigonometric ratio to write an equation. Solve the equation for the missing value.

Similarity, Right Triangles, and Trigonometry
Set 3: Sine, Cosine, and Tangent Ratios, and Angles of Elevation and Depression

Instruction

Possible Misunderstandings/Mistakes

- Incorrectly identifying the opposite side and adjacent side when writing trigonometric ratios
- Incorrectly writing the trigonometric ratios
- Mixing up the angle of elevation with the angle of depression
- Mixing up when to use the trigonometric function and when to use the inverse of the function when finding missing side lengths or angles

Similarity, Right Triangles, and Trigonometry

Set 3: Sine, Cosine, and Tangent Ratios, and Angles of Elevation and Depression

Station 1

You will be given a graphing calculator, a ruler, and a protractor. Work together to answer the questions.

In the space below, construct a right triangle with legs that are 3 inches and 4 inches in length, and a hypotenuse that is 5 inches in length. Label the angle opposite the shorter leg as α (alpha). Label the angle opposite the longer leg as β (beta).

The triangle is a right triangle; you can write trigonometric ratios to represent the values of α and β. Review the trigonometric functions:

$$\sin \theta = \frac{\text{opposite}}{\text{hypotenuse}}; \cos \theta = \frac{\text{adjacent}}{\text{hypotenuse}}; \tan \theta = \frac{\text{opposite}}{\text{adjacent}}$$

1. Find sin(α).

2. Find cos(α).

3. Find tan(α).

4. Find sin(β).

5. Find cos(β).

6. Find tan(β).

continued

Similarity, Right Triangles, and Trigonometry
Set 3: Sine, Cosine, and Tangent Ratios, and Angles of Elevation and Depression

You can use the trigonometric ratios and your graphing calculator to find the measure of angle α.

For example, let's say $\cos(\alpha) = \dfrac{3}{4}$.

Follow these steps to find the measure of α:

    1: Hit the "2nd" key. Hit the "COS" key.

    2: Type in (3/4) or 0.75. Hit "Enter."

    Answer = 41.4°

7. Find the measure of α and β in the triangle you drew.

Similarity, Right Triangles, and Trigonometry

Set 3: Sine, Cosine, and Tangent Ratios, and Angles of Elevation and Depression

Station 2

You will be given a graphing calculator. Work as a group to answer the questions.

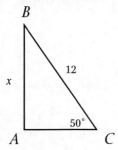

1. How can you write the sine ratio for the 50° angle in the triangle above?

2. Find the value of *x* using this sine ratio. (*Hint:* First, use your calculator to find sin 50°.) Show your work and answer in the space below.

3. In the space below, construct a 45°–45°–90° triangle. Use your knowledge of the 45°–45°–90° right triangles and trigonometric ratios to prove that $\sin\theta = \dfrac{\text{opposite}}{\text{hypotenuse}}$.

continued

Similarity, Right Triangles, and Trigonometry

Set 3: Sine, Cosine, and Tangent Ratios, and Angles of Elevation and Depression

For problems 4–6, find the value of x. Round your answer to the nearest hundredth.

4.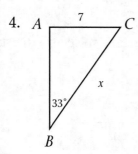

5.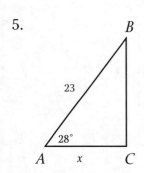

6.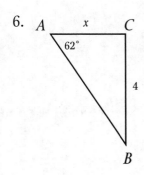

NAME: _____

Similarity, Right Triangles, and Trigonometry
Set 3: Sine, Cosine, and Tangent Ratios, and Angles of Elevation and Depression

Station 3

You will be given three drinking straws, a graphing calculator, scissors, tape, and a ruler. Work as a group to solve the following problem.

Construct a small right triangle out of the three drinking straws. Use the tape to secure the straws together. Then trim off the excess straws not in the right triangle.

1. What are the lengths of the legs of the triangle? _____

2. What is the length of the hypotenuse of the triangle? _____

Imagine that one of the straws in your triangle is a tree. This tree makes a 90° angle with the ground. The tree casts a shadow on the ground. Tape your triangle to your table so it remains upright to represent this situation.

3. What is the height of the straw that represents your tree? _____
   Is this a leg or hypotenuse of the triangle? Explain your answer.

4. What is the length of the straw that represents the shadow cast from the top of the tree to the ground? _____
   Is this a leg or hypotenuse of the triangle? Explain your answer.

5. What is the length of the straw that represents the distance between the tree and the point on the ground where the shadow ends? _____
   Is this a leg or hypotenuse of the triangle? Explain your answer.

6. What is the angle of elevation from the end of the shadow to the top of the tree? Show your work and answer on a separate sheet of paper. (*Hint:* Use trigonometric ratios.)

Similarity, Right Triangles, and Trigonometry
Set 3: Sine, Cosine, and Tangent Ratios, and Angles of Elevation and Depression

Station 4

You will be given a tape measure and tape. Work together to solve the following problem.

Have two students stand 6 feet away from each other. Have the first student stand on a chair. Model the diagram below.

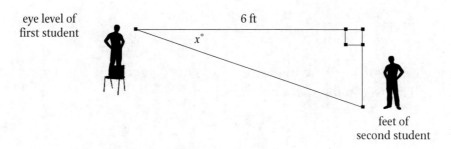

1. Measure the distance between the eye level of the first student and the feet of the second student. What is this distance?

   _____

2. Find the angle of depression, represented by x in the diagram, using trigonometric ratios. Show your work and answer in the space below.

3. The angle represented by x in the diagram represents the angle of depression. Why do you think the angle has this name?

continued

Similarity, Right Triangles, and Trigonometry
Set 3: Sine, Cosine, and Tangent Ratios, and Angles of Elevation and Depression

Repeat the procedure used in problems 1 and 2 by having two students stand 3 feet apart.

4. Measure the distance between the eye level of the first student and the feet of the second student. What is this distance?

   _____

5. Find the angle of depression, represented by x in the diagram, using trigonometric ratios. Show your work and answer in the space below.

6. How does your answer in problem 2 compare with your answer in problem 5?

Circles

Set 1: Circumference, Angles, Arcs, Chords, and Inscribed Angles

Instruction

Goal: To provide opportunities for students to develop concepts and skills related to circumference, arc length, central angles, chords, and inscribed angles

Common Core Standards

Congruence

Experiment with transformations in the plane.

G-CO.1. Know precise definitions of angle, circle, perpendicular line, parallel line, and line segment, based on the undefined notions of point, line, distance along a line, and distance around a circular arc.

Circles

Understand and apply theorems about circles.

G-C.2. Identify and describe relationships among inscribed angles, radii, and chords. Include the relationship between central, inscribed, and circumscribed angles; inscribed angles on a diameter are right angles; the radius of a circle is perpendicular to the tangent where the radius intersects the circle.

Find arc lengths and areas of sectors of circles.

G-C.5. Derive using similarity the fact that the length of the arc intercepted by an angle is proportional to the radius, and define the radian measure of the angle as the constant of proportionality; derive the formula for the area of a sector.

Student Activities Overview and Answer Key

Station 1

Students will be given a plastic coffee can lid, a tape measure, a black marker, a compass, a ruler, and white paper. Students will measure the radius, diameter, and circumference of the coffee can lid. They will derive the relationships between circumference and radius and circumference and diameter. Then they will solve a real-world problem using circumference.

Answers

1. Circumference; answers will vary.

2. Verify the circle on students' papers.

3. Answers will vary; answers will vary; $d = 2r$

4. the ratio of the circumference of a circle to its diameter, which may be approximated by 3.14

Circles
Set 1: Circumference, Angles, Arcs, Chords, and Inscribed Angles

Instruction

5. Answers will vary; yes, this calculation finds the circumference.

6. Answers will vary; yes, this calculation finds the circumference.

7. Yes, this calculation finds the circumference.

8. $C = 2\pi r$; $C = \pi d$

9. $C = \pi d = \pi(20) = 62.8$ ft

10. $C = 2\pi r$; $6.28 = 2\pi r$; $6.28 = 6.28r$; $r = 1$ inch

Station 2

Students will be given a compass, a protractor, a red marker, a calculator, and a ruler. They will construct a circle, radii, a central angle, and an arc. Students will derive the relationship between the arc and the circumference. Then they will find the length of the arc.

Answers

1. 6.28 inches

2. $\frac{1}{4}$; $\frac{90°}{360°} = \frac{1}{4}$

3. $\frac{1}{4}(6.28) = 1.57$ inches

4. length of arc = $\left(\frac{\text{central angle}}{360°}\right)(\text{circumference})$

5. 2.62 cm

6. 8.85 in

7. 2.36 m

8. 38.47 ft

Station 3

Students will be given white computer paper, a compass, a ruler, a red marker, a blue marker, and a calculator. Students will construct circles, radii, and chords. They will derive the relationship between chords, arcs, and the triangles created by chords and radii. Then they will solve a real-world problem.

Circles
Set 1: Circumference, Angles, Arcs, Chords, and Inscribed Angles

Instruction

Answers

1. 9.42 in
2. 1.05 in; 1.05 in
3. 1.05 in; 1.05 in
4. length of arc = length of chord
5. $\triangle PAB$ and $\triangle PCD$ are congruent because of the SAS theorem. Both triangles have congruent sides (the radii) and the included corresponding angles are congruent since they each measure 40 degrees.
6. $m\overarc{AB}$ and $m\overarc{CD}$ are congruent because the central angles are the same measure.
7. $C = \pi d = \pi(10) = 31.4$ in. Cut the pie into 6 slices by dividing 31.4 by 6 to find the arc length of one slice: 31.4/6 = 5.23 in. Chord length is equal to arc length.

Station

Students will be given white computer paper, a compass, a protractor, a ruler, and a calculator. They will construct circles, inscribed angles, and central angles. Students will find the degree measure of the inscribed angle and its intercepted arc. Then they will derive the relationship between the inscribed angle and its intercepted arc.

Answers

1. Answers will vary.
2. Answers will vary. (Should be twice the answer in problem 1.)
3. Answers will vary.
4. Answers will vary. (Should be twice the answer in problem 3.)
5. The measure of an inscribed angle is equal to one-half the degree measure of its intercepted arc.

Materials List/Setup

Station 1   plastic coffee can lid; tape measure; black marker; compass; ruler; white paper
Station 2   compass; protractor; red marker; calculator; ruler
Station 3   white computer paper; compass; ruler; red marker; blue marker; calculator
Station 4   white computer paper; compass; protractor; ruler; calculator

Circles
Set 1: Circumference, Angles, Arcs, Chords, and Inscribed Angles

Instruction

Discussion Guide

To support students in reflecting on the activities and to gather some formative information about student learning, use the following prompts to facilitate a class discussion to "debrief" the station activities.

Prompts/Questions

1. How do you find the circumference of a circle given the radius? Given the diameter?
2. In a circle, how do you find the length of an arc given its central angle?
3. In a circle, what is the relationship between arcs and chords?
4. In a circle, what is the relationship between an inscribed angle and its intercepted arc?

Think, Pair, Share

Have students jot down their own responses to questions, then discuss with a partner (who was not in the same station group), and then discuss as a whole class.

Suggested Appropriate Responses

1. $C = 2\pi r$; $C = \pi d$
2. length of arc $= \left(\dfrac{\text{central angle}}{360°} \right) (\text{circumference})$
3. length of arc = length of chord
4. The measure of an inscribed angle is equal to one-half the degree measure of its intercepted arc.

Possible Misunderstandings/Mistakes

- Not understanding the difference between a radius and a chord
- Not finding the measure of the correct arc
- Not having the vertex of an inscribed angle on the circle

NAME:

Circles
Set 1: Circumference, Angles, Arcs, Chords, and Inscribed Angles

Station 1

You will be given a plastic coffee can lid, a tape measure, a black marker, a compass, a ruler, and white paper.

1. As a group, use the black marker to mark a starting point on your coffee lid.

   Roll the coffee can lid along the tape measure so you can measure the distance around the edge of the coffee can lid.

   What is the mathematical name for this distance?

   What is the distance around the edge of the lid in inches?

   Repeat this measurement three more times to verify your answer.

2. Trace the coffee can lid on the white paper.
   Use the ruler and compass to find the center of the circle.

3. What is the radius of the circle?

   What is the diameter of the circle?

   How does the radius relate to the diameter?

4. What is π?

continued

Circles
Set 1: Circumference, Angles, Arcs, Chords, and Inscribed Angles

5. What is π times twice the radius of your circle?

   Does this match your answer in problem 1? Why or why not?

6. What is π times the diameter of your circle?

   Does this match your answer in problem 1? Why or why not?

7. Do your answers for problems 5 and 6 match? Why or why not?

8. Based on your observations in problems 1–7, what is the formula for the circumference of a circle written in terms of the radius?

   What is the formula for the circumference of a circle written in terms of the diameter?

continued

Circles
Set 1: Circumference, Angles, Arcs, Chords, and Inscribed Angles

9. Larry installed a circular pool in his backyard. The pool has a diameter of 20 feet. What is the circumference of the pool? Show your work in the space below.

10. Lisa is running for class president and passed out buttons that each have a circumference of 6.28 inches. What is the radius of each button? Show your work in the space below.

Circles

Set 1: Circumference, Angles, Arcs, Chords, and Inscribed Angles

Station 2

You will be given a compass, a protractor, a red marker, a calculator, and a ruler. Work as a group to construct the circles and answer the questions.

1. In the space below, construct a circle with a diameter of 2 inches. Label the center of the circle as point C.

What is the circumference of the circle?

continued

Circles
Set 1: Circumference, Angles, Arcs, Chords, and Inscribed Angles

2. On the circle, construct a horizontal radius. Use the protractor to create a vertical radius that creates a 90° angle with your horizontal radius.

   Use the red marker to highlight the arc of the circle between the endpoints of these radii.

   What fraction of the circle's circumference is this arc?

   What is the ratio of the central angle you created between the two radii and the total angle measure of the circle? Justify your answer.

3. How can you use the circumference of the circle and the ratio of the central angle to the total angle measure of the circle to find the length of the arc? Explain your answer.

4. In general, what method can you use to find the length of an arc given the central angle?

continued

Circles
Set 1: Circumference, Angles, Arcs, Chords, and Inscribed Angles

For problems 5–8, find the length of the arc using the given information.

5. Circle with radius 5 cm and central angle of 30°

6. Circle with diameter 7 inches and central angle of 145°

7. Circle with radius 0.5 meters and central angle of 270°

8. Circle with diameter 14 feet and central angle of 315°

NAME:

Circles
Set 1: Circumference, Angles, Arcs, Chords, and Inscribed Angles

Station 3

You will be given white computer paper, a compass, a ruler, a red marker, a blue marker, and a calculator. Work as a group to construct the circles and answer the questions.

On the paper, construct a circle that has a diameter of 3 inches. Mark the center of the circle as point P.

1. What is the circumference of the circle?

2. Construct two radii, $\overline{PA}$ and $\overline{PB}$, that create a 40° angle. Draw chord $\overline{AB}$.

   What is the length of chord $\overline{AB}$?

   What is the length of $\overset{\frown}{AB}$? (*Hint:* Use $\left(\dfrac{\text{central angle}}{360°}\right)(\text{circumference})$.)

3. Construct two radii, $\overline{PC}$ and $\overline{PD}$, that also create a 40° angle. Draw chord $\overline{CD}$.

   What is the length of chord $\overline{CD}$?

   What is the length of $\overset{\frown}{CD}$?

continued

Circles
Set 1: Circumference, Angles, Arcs, Chords, and Inscribed Angles

4. Based on your observations in problems 1–3, what is the relationship between chords and arcs?

5. What is the relationship between $\triangle PAB$ and $\triangle PCD$? Explain your answer.

6. What is the relationship between $m\overarc{AB}$ and $m\overarc{CD}$? Explain your answer.

7. An apple pie 10 inches in diameter is cut into 6 equal size slices. What is the length of the chord for each slice of pie? Show your work and answer in the space below.

Circles
Set 1: Circumference, Angles, Arcs, Chords, and Inscribed Angles

Station 4

You will be given white computer paper, a compass, a protractor, a ruler, and a calculator. Work as a group to construct the circles and answer the questions.

On the white computer paper, construct a circle with a radius of 0.75 inches.

1. Plot a point on the circle. Label this point P.

    Construct two chords, $\overline{PA}$ and $\overline{PB}$, to create an inscribed angle.

    What is the measure of inscribed angle $\angle APB$?

2. Label the center of the circle as point C.

    Construct two radii, $\overline{CA}$ and $\overline{CB}$.

    In the space below, find $m\overset{\frown}{AB}$.

continued

Circles
Set 1: Circumference, Angles, Arcs, Chords, and Inscribed Angles

On the white paper, construct a new circle with a radius of 2 inches.

3. Plot a point on the circle. Label this point P.

   Construct two chords, $\overline{PA}$ and $\overline{PB}$, to create an inscribed angle.

   What is the measure of the inscribed angle $\angle APB$?

4. Label the center of the circle as point C.

   Construct two radii, $\overline{CA}$ and $\overline{CB}$.

   In the space below, find $m\overset{\frown}{AB}$.

5. Based on your observations in problems 1–4, what is the relationship between the measure of an inscribed angle and its intercepted arc?

Circles

Set 2: Special Segments, Angle Measurements, and Equations of Circles

Instruction

Goal: To provide opportunities for students to develop concepts and skills related to special segments, secants, tangents, angle measurements, and equations of circles

Common Core Standards

Congruence

Experiment with transformations in the plane.

G-CO.1. Know precise definitions of angle, circle, perpendicular line, parallel line, and line segment, based on the undefined notions of point, line, distance along a line, and distance around a circular arc.

Circles

Understand and apply theorems about circles.

G-C.2. Identify and describe relationships among inscribed angles, radii, and chords. Include the relationship between central, inscribed, and circumscribed angles; inscribed angles on a diameter are right angles; the radius of a circle is perpendicular to the tangent where the radius intersects the circle.

G-C.4. (+) Construct a tangent line from a point outside a given circle to the circle.

Find arc lengths and areas of sectors of circles.

G-C.5. Derive using similarity the fact that the length of the arc intercepted by an angle is proportional to the radius, and define the radian measure of the angle as the constant of proportionality; derive the formula for the area of a sector.

Translate between the geometric description and the equation for a conic section.

G-C.1. Derive the equation of a circle of given center and radius using the Pythagorean theorem; complete the square to find the center and radius of a circle given by an equation.

Use coordinates to prove simple geometric theorems algebraically.

G-GPE.4. Use coordinates to prove simple geometric theorems algebraically.

Student Activities Overview and Answer Key

Station 1

Students will be given a ruler and calculator. They will work together to construct secants that intersect outside a circle. They will derive the relationship between the secants using the secant segment and external portions.

Circles
Set 2: Special Segments, Angle Measurements, and Equations of Circles

Instruction

Answers
1. The secants intersect at a point outside the circle.
2. $FA \approx 1.2$ inches
3. $FB \approx 2.9$ inches
4. $FC \approx 1.3$ inches
5. $FD \approx 2.7$ inches
6. $FA \times FB \approx 3.5$ square inches
7. $FC \times FD \approx 3.5$ square inches
8. If two secants intersect outside a circle, then the product of the secant segment with its external portion equals the product of the other secant segment with its external portion.

Station 2

Students will be given a ruler, a compass, a protractor, and a calculator. Students will construct a secant and tangent on a circle. They will measure the angle created by the secant and tangent from a point drawn outside the circle. They will find the measure of the intercepted arcs. Then they will compare the angle and measure of intercepted arcs.

Answers
1.
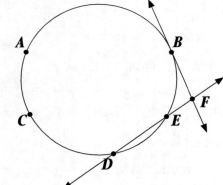

2. 105°
3. 260°
4. 50°
5. The angle formed equals half the difference of the intercepted arcs.

Circles
Set 2: Special Segments, Angle Measurements, and Equations of Circles

Instruction

Station 3

Students will be given white computer paper, a ruler, a compass, a protractor, and a calculator. Students will construct a circle and tangents that intersect at a point outside the circle. They will derive the relationship between the angle formed by the tangents and the intercepted arcs.

Answers

1. Answers will vary. Sample answer: outside the circle

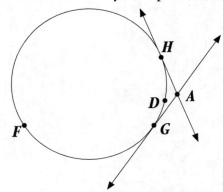

2. Answers will vary. Sample answer: 115°
3. Answers will vary. Possible answer based on answer to question 2: 295°
4. Answers will vary. Possible answer based on answer to question 2: 65°
5. The measure of an angle formed by two tangents drawn from a point outside the circle is half the difference of the intercepted arcs.

Station 4

Students will be given graph paper, a ruler, and a compass. Students will construct circles in the coordinate plane. They will find the equation of a circle given the center and radius, and state the center and radius of the circle when the equation of the circle is given in center-radius form.

Answers

1. $h = 0; k = 0; x^2 + y^2 = 25$
2. $h = 10; k = -4; (x - 10)^2 + (y + 4)^2 = 25$
3. Answers will vary. Possible answer: The equation of the circle $x^2 + y^2 = r^2$ is based on a circle with a center at the origin (0, 0). Both h and k give the center of the circle in relationship to the center of the circle at $x^2 + y^2 = r^2$.

Circles
Set 2: Special Segments, Angle Measurements, and Equations of Circles

Instruction

4. $(-3, 6)$; $r = 5$; Sample graph:

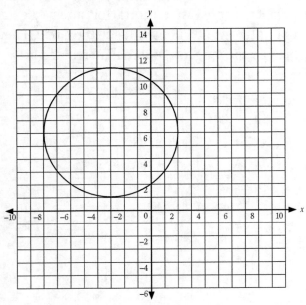

5. $(1, 2)$; $r = 6$; Sample graph:

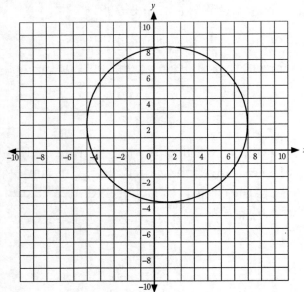

Materials List/Setup

Station 1   ruler; calculator

Station 2   ruler; compass; protractor; calculator

Station 3   white computer paper; ruler; compass; protractor; calculator

Station 4   graph paper; ruler; compass

Circles
Set 2: Special Segments, Angle Measurements, and Equations of Circles

Instruction

Discussion Guide

To support students in reflecting on the activities and to gather formative information about student learning, use the following prompts to facilitate a class discussion to "debrief" the station activities.

Prompts/Questions

1. What is the relationship between two secants that intersect outside of a circle?
2. What is the relationship between the measure of the angle created by the secant and tangent drawn from a point outside the circle and the intercepted arcs?
3. What is the relationship between the measure of the angle created by two tangents drawn from a point outside the circle and the intercepted arcs?
4. How do you write the equation for a circle given the coordinates of the center point and radius?

Think, Pair, Share

Have students jot down their own responses to questions, then discuss with a partner (who was not in the same station group), and then discuss as a whole class.

Suggested Appropriate Responses

1. If two secants intersect outside a circle, then the product of the secant segment with its external portion equals the product of the other secant segment with its external portion.
2. The angle formed equals half the difference of the intercepted arcs.
3. The measure of an angle formed by two tangents drawn from a point outside the circle is half the difference of the intercepted arcs.
4. Use $(x - h)^2 + (y - k)^2 = r^2$.

Possible Misunderstandings/Mistakes

- Not measuring the correct segments when comparing secants that intersect at a point outside the circle
- Not calculating the measure of intercepted arcs correctly
- Not identifying the intercepted arcs correctly
- Not changing the subtraction sign to a plus sign when a coordinate of the center of the circle is negative

Circles

Set 2: Special Segments, Angle Measurements, and Equations of Circles

Station 1

You will be given a ruler and calculator. Work as a group to answer the questions.

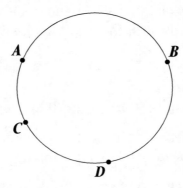

On the circle above, construct a secant through points C and D. Construct a secant through points A and B.

1. Where do the secants intersect?

   Label this intersection as point F.

2. What is the length of $\overline{FA}$?

3. What is the length of $\overline{FB}$?

4. What is the length of $\overline{FC}$?

5. What is the length of $\overline{FD}$?

6. What is $\overline{FA} \times \overline{FB}$?

7. What is $\overline{FC} \times \overline{FD}$?

8. Based on your observations in problems 1–7, what is the relationship between two secants that intersect outside of a circle?

Circles

Set 2: Special Segments, Angle Measurements, and Equations of Circles

Station 2

You will be given a ruler, a compass, a protractor, and a calculator. Work as a group to answer the questions.

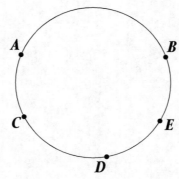

On the circle above, construct a secant through points *D* and *E*. Construct a tangent through point *B*.

1. Do the secant and tangent intersect inside or outside the circle?

2. Label the angle created by the secant and tangent as $\angle F$.

   What is the measure of $\angle F$?

3. What is $m\overset{\frown}{DCB}$? Show your work in the space below.

continued

Circles
Set 2: Special Segments, Angle Measurements, and Equations of Circles

4. What is $m\overset{\frown}{BE}$? Show your work in the space below.

5. Based on your observations in problems 1–4, what is the relationship between the measure of the angle created by the secant and tangent drawn from a point outside the circle and the intercepted arcs?

Circles
Set 2: Special Segments, Angle Measurements, and Equations of Circles

Station 3

You will be given white computer paper, a ruler, a compass, a protractor, and a calculator. Work as a group to construct the circles and answer the questions.

- Construct a circle on the computer paper.
- Construct two points on the right edge of the circle. Label the points as *H* and *G*.
- Construct a point on the right edge of the circle between points *H* and *G*. Label this point *D*.
- Construct a point on the left edge of the circle between points *H* and *G*. Label this point *F*.
- Construct tangent lines through points *H* and *G*.

1. Where do the tangent lines intersect?

   Label this as point *A*.

2. What is the measure of ∠*A*?

3. What is $m\overarc{HFG}$? Show your work and answer in the space below.

Circles
Set 2: Special Segments, Angle Measurements, and Equations of Circles

4. What is $m\overset{\frown}{GDH}$? Show your work and answer in the space below.

5. What is the relationship between the measure of the angle formed by the point of intersection of the two tangent lines and the intercepted arcs? Show your work and answer in the space below.

Circles
Set 2: Special Segments, Angle Measurements, and Equations of Circles

Station 4

You will be given graph paper, a ruler, and a compass. Work as a group to construct the circles and answer the questions.

The equation of a circle is $(x - h)^2 + (y - k)^2 = r^2$.

1. On your graph paper, construct a circle with its center at (0, 0) and a radius of 5 units.

    What is the value of h for this circle?

    What is the value of k for this circle?

    What is the equation of this circle?

2. On your graph paper, shift the circle in problem 1 down 4 units and right 10 units.

    What is the value of h for this circle?

    What is the value of k for this circle?

    What is the equation of this circle?

continued

Circles
Set 2: Special Segments, Angle Measurements, and Equations of Circles

3. Why do you think h and k are used in the equation of a circle?

4. A circle has the equation $(x+3)^2 + (y-6)^2 = 25$.

   What is the center of the circle? Explain your answer.

   What is the radius of the circle? Explain your answer.

   Graph the circle on your graph paper to justify your answers.

5. A circle has the equation $(x-1)^2 + (y-2)^2 = 36$.

   What is the center of the circle? Explain your answer.

   What is the radius of the circle? Explain your answer.

   Graph the circle on your graph paper to justify your answers.

Circles

Set 3: Circumcenter, Incenter, Orthocenter, and Centroid

Instruction

Goal: To provide opportunities for students to develop concepts and skills related to the circumcenter, incenter, orthocenter, and centroid of a circle

Common Core Standards

Congruence

Experiment with transformations in the plane.

G-CO.1. Know precise definitions of angle, circle, perpendicular line, parallel line, and line segment, based on the undefined notions of point, line, distance along a line, and distance around a circular arc.

Make geometric constructions.

G-CO.12. Make formal geometric constructions with a variety of tools and methods (compass and straightedge, string, reflective devices, paper folding, dynamic geometric software, etc.).

G-CO.13. Construct an equilateral triangle, a square, and a regular hexagon inscribed in a circle.

Circles

Understand and apply theorems about circles.

G-C.3. Construct the inscribed and circumscribed circles of a triangle, and prove properties of angles for a quadrilateral inscribed in a circle.

Student Activities Overview and Answer Key

Station 1

Students will be given a ruler, a protractor, and a compass. They will construct a square and circumscribe a circle around the square. Then they will construct the perpendicular bisectors of the square. They will realize that the circumcenter is the intersection of the perpendicular bisectors, which is also the center of the circle. They will repeat this process for an equilateral triangle.

Circles
Set 3: Circumcenter, Incenter, Orthocenter, and Centroid

Instruction

Answers

1–2.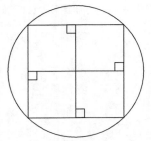

The circle circumscribes the square with the intersection of the perpendicular bisectors as the center of the circle.

3.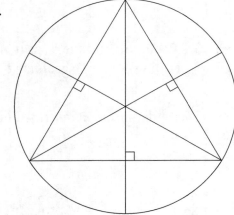

4. circumcenter

5. The circle circumscribes the triangle with each vertex being on the edge of the circle and the center being the intersection of the perpendicular bisectors.

6. The circumcenter is the center of the circle that circumscribes the polygon.

Station 2

Students will be given a ruler, a compass, and a protractor. They will construct the angle bisectors of an equilateral triangle. Then they will inscribe a circle in the triangle. They will realize that the incenter is the intersection of the angle bisectors, which is also the center of the circle.

Circles
Set 3: Circumcenter, Incenter, Orthocenter, and Centroid

Instruction

Answers

1., 3., 4.

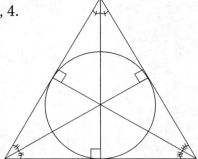

2. incenter

5. The circle is inscribed in the triangle.

6. The incenter is the center of an inscribed circle and the point where the angle bisectors of a regular polygon intersect.

Station 3

Students will be given a ruler, a protractor, and a compass. Students will construct the altitudes of an acute triangle to find the orthocenter. Then they will construct the altitudes of a right triangle and an obtuse triangle and discover that the orthocenter can lie inside, on, or outside the triangle.

Answers

1.

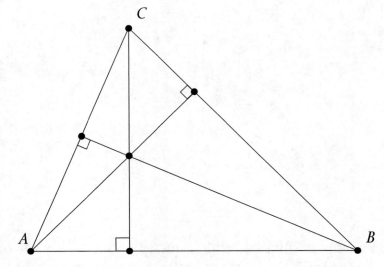

2. orthocenter

3. acute

Circles
Set 3: Circumcenter, Incenter, Orthocenter, and Centroid

Instruction

4.

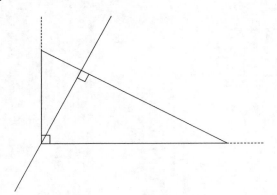

5. on the right angle of the triangle
6. right

7.

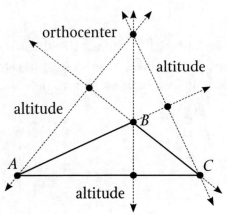

8. outside the triangle
9. obtuse
10. The orthocenter can lie inside, on, or outside the triangle.

Station 4

Students will be given notecards, scissors, a compass, a ruler, and a protractor. They will construct the centroid of three triangles. They will relate the centroid to the center of gravity or the balancing point.

Circles
Set 3: Circumcenter, Incenter, Orthocenter, and Centroid

Instruction

Answers

1. Triangles may vary. Be sure all angles are acute in the triangle. Sample triangle with medians:

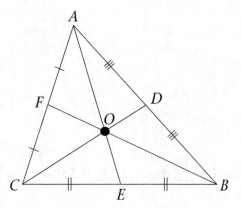

2. centroid

3. Triangles may vary. Be sure there is a right angle in the triangle. Sample triangle with medians:

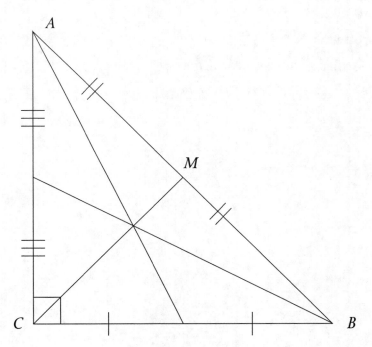

Circles
Set 3: Circumcenter, Incenter, Orthocenter, and Centroid

Instruction

4. Triangles may vary. Be sure there is an obtuse angle in the triangle. Sample triangle with medians:

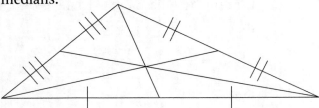

5. at the intersection of the medians
6. centroid
7. The centroid is the center of gravity and is formed by the intersection of the medians.

Materials List/Setup

Station 1 ruler; compass; protractor
Station 2 ruler; compass; protractor
Station 3 ruler; compass; protractor
Station 4 notecards; scissors; ruler; compass; protractor

Circles
Set 3: Circumcenter, Incenter, Orthocenter, and Centroid

Instruction

Discussion Guide

To support students in reflecting on the activities and to gather some formative information about student learning, use the following prompts to facilitate a class discussion to "debrief" the station activities.

Prompts/Questions

1. What is the circumcenter of a circle circumscribed about a regular polygon?
2. What is the incenter of a circle inscribed in a regular polygon?
3. What is the orthocenter?
4. What is the centroid of a circle?

Think, Pair, Share

Have students jot down their own responses to questions, then discuss with a partner (who was not in their station group), and then discuss as a whole class.

Suggested Appropriate Responses

1. It is the intersection of the perpendicular bisectors of the circumscribed polygons.
2. It is the intersection of the angle bisectors of the inscribed polygons.
3. It is the intersection of the altitudes of a triangle.
4. The centroid is the intersection of the medians and the center of gravity for the figure.

Possible Misunderstandings/Mistakes

- Not realizing that the circumcenter is found in circumscribed circles
- Not realizing that the incenter is found in inscribed circles
- Incorrectly constructing perpendicular bisectors, angle bisectors, and altitudes

Circles
Set 3: Circumcenter, Incenter, Orthocenter, and Centroid

Station 1

At this station, you will find a ruler, a compass, and a protractor. Work as a group to answer the questions.

1. In the space below, construct the perpendicular bisectors of the sides of the square.

2. Draw a circle using the intersection of the perpendicular bisectors as the center of the circle and one of the vertices of the square as the radius. What do you notice about the circle in relation to the square?

continued

Circles
Set 3: Circumcenter, Incenter, Orthocenter, and Centroid

3. In the space below, construct the perpendicular bisectors of the sides of the triangles.

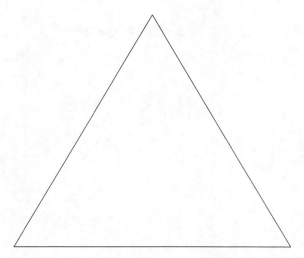

4. What is the name of the point of intersection of the perpendicular bisectors of a triangle?

5. Draw a circle using the intersection of the perpendicular bisectors as the center of the circle and one of the vertices of the triangle as the radius. What do you notice about the circle in relationship to the triangle?

6. Based on your observations in problems 1–4, what is the definition of the circumcenter of a circumscribed polygon?

NAME: _____

Circles
Set 3: Circumcenter, Incenter, Orthocenter, and Centroid

Station 2

At this station, you will find a ruler, a compass, and a protractor. Work as a group to answer the questions.

1. Using the triangle below, bisect each angle of the triangle.

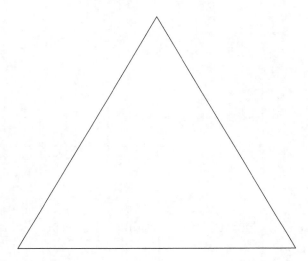

2. What is the name of the intersection point of the angle bisectors of a triangle?

3. Construct a perpendicular line from the intersection of the angle bisectors of the triangle to one of the sides of the triangle.

4. Draw a circle using the intersection of the angle bisectors of the triangle as the center and the intersection of the perpendicular line constructed in problem 2 with the side of the triangle as the radius.

continued

Circles
Set 3: Circumcenter, Incenter, Orthocenter, and Centroid

5. What do you notice about the circle in relation to the triangle?

6. Based on your observations in problems 1–5, what is the definition of the incenter of an inscribed polygon?

NAME:

Circles
Set 3: Circumcenter, Incenter, Orthocenter, and Centroid

Station 3

At this station, you will find a ruler, a compass, and a protractor. Work as a group to answer the questions.

1. On the triangle below, construct the altitudes.

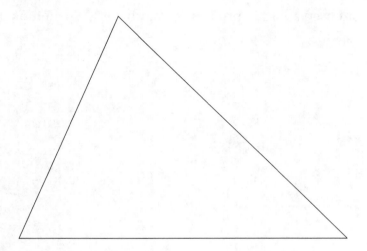

2. What is the name of the point of intersection of the altitudes of a triangle?
   _____

3. Based on the angles, what type of triangle is this? _____

NAME:

Circles
Set 3: Circumcenter, Incenter, Orthocenter, and Centroid

4. On the triangle below, construct the altitudes.

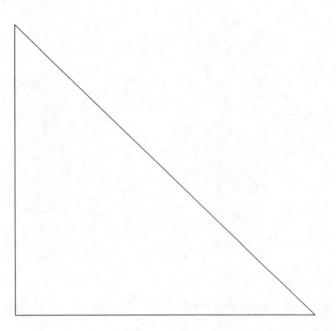

5. Where does the intersection of the altitudes occur? _____

6. Based on the angles, what type of triangle is this? _____

continued

Circles
Set 3: Circumcenter, Incenter, Orthocenter, and Centroid

7. On the triangle below, construct the altitudes. Remember that you might need to extend the leg(s) of the triangle to create the altitude(s).

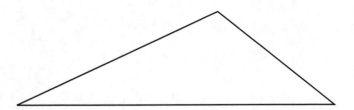

8. Where does the intersection of the altitudes occur? _____

9. Based on the angles, what type of triangle is this? _____

10. Based on your answers to problems 1–9, what conclusion can you draw about the location of the intersection of the altitudes of a triangle?

NAME:

Circles
Set 3: Circumcenter, Incenter, Orthocenter, and Centroid

Station 4

At this station, you will find notecards, scissors, a compass, a ruler, and a protractor. Work together to construct the medians of the triangle and answer the questions.

1. Draw an acute triangle on one of the notecards and then construct the medians of the triangle.

2. Identify the intersection of the medians with a point. What is this point called?

3. Draw a right triangle on one of the notecards and then construct the medians of the triangle.

4. Draw an obtuse triangle on one of the notecards and then construct the medians of the triangle.

5. Cut out each of the triangles and try balancing each one of them on the tip of your finger. Where does the balancing point occur?

6. What is the intersection point of the medians called?

7. What conclusion can you draw about the intersection point of the medians of a triangle?

Expressing Geometric Properties with Equations

Set 1: Parallel Lines, Slopes, and Equations

Instruction

Goal: To provide opportunities for students to develop concepts and skills related to using coordinate geometry to find slopes, parallel lines, perpendicular lines, and equations of lines

Common Core Standards
Congruence

Experiment with transformations in the plane.

G-CO.1. Know precise definitions of angle, circle, perpendicular line, parallel line, and line segment, based on the undefined notions of point, line, distance along a line, and distance around a circular arc.

Expressing Geometric Properties with Equations

Use coordinates to prove simple geometric theorems algebraically.

G-GPE.4. Use coordinates to prove simple geometric theorems algebraically.

G-GPE.5. Prove the slope criteria for parallel and perpendicular lines and use them to solve geometric problems (e.g., find the equation of a line parallel or perpendicular to a given line that passes through a given point).

Student Activities Overview and Answer Key

Station 1

Students will be given graph paper, a ruler, a red marker, and a blue marker. Students will use the graph paper and ruler to construct straight lines. Then they will use the red and blue markers to depict the rise and run of the slope of the lines. They will determine whether two lines are parallel by constructing the lines, or given two points on the line, or given the slope of the lines.

Answers

1. $m = 2/3$
2. $m = 2/3$
3. $m = 4/5$
4. Line 1 and Line 2 because they have the same slope of 2/3.
5. No; the slopes are not equal.
6. Yes; the slopes are equal.
7. Yes; the slopes are equal.
8. No; the slopes are not equal.

Expressing Geometric Properties with Equations
Set 1: Parallel Lines, Slopes, and Equations

Instruction

Station 2

Students will be given a protractor. Students will identify corresponding angles of two lines cut by a transversal. Then they will use the protractor to verify their answer. They will determine whether two lines are parallel using corresponding, supplementary, and vertical angles.

Answers

1. 1 and 3; 2 and 4
2. $m\angle 1 = 130°$, $m\angle 2 = 50°$, $m\angle 3 = 130°$, $m\angle 4 = 50°$
3. 1 and 3; 2 and 4
4. Yes, because the corresponding angles have equal measure.
5. Answers will vary. Possible answer: The top angle measuring 115° has a supplementary angle of 65°. The corresponding angle is 65°, which is supplementary to the bottom angle measuring 115°. Because of this, the lines are parallel.
6. Answers will vary. Possible answer: The top angle measuring 115° and the bottom angles measuring 115° both have a supplementary angle of 65°. This means that the angle above the upper 115° and the angle to the left of the lower 115° are both 65°. These are also corresponding angles, and since they have the same measure, then the lines are parallel.

Station 3

Students will be given uncooked spaghetti noodles, graph paper, and a ruler. Students will use the graph paper and ruler to construct straight lines. They will use the spaghetti noodles to model parallel lines. They will find the slope of the lines to determine if lines are parallel. Then they will construct a line perpendicular to the two lines to determine if the lines are parallel. They also will use their knowledge of corresponding angles to determine if the two lines are parallel.

Answers

1.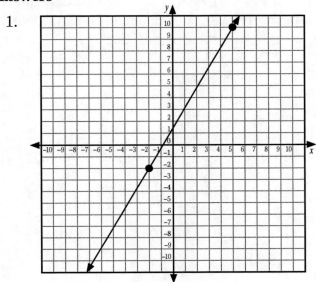

Expressing Geometric Properties with Equations
Set 1: Parallel Lines, Slopes, and Equations

Instruction

2.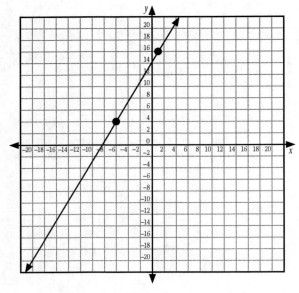

3. Find the slopes of the lines. Slope = 12/7 for both lines, so, yes, the lines are parallel.

4. Slope = –7/12. Yes, because the slope is the opposite reciprocal. Yes, because the slope is the opposite reciprocal.

5. 90°

6. Corresponding angles are equal at 90°, so the lines are parallel.

7. perpendicular; parallel

Station 4

Students will be given graph paper and a ruler. Students will construct a straight line and find its slope. Then they will use the slope and point-slope form of a linear equation to construct two lines parallel to the original line. They will write the equations for these parallel lines.

Expressing Geometric Properties with Equations
Set 1: Parallel Lines, Slopes, and Equations

Instruction

Answers

1.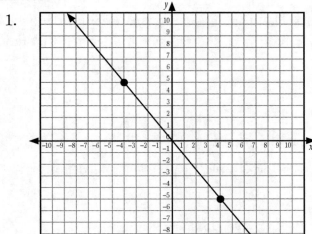

2. $m = -\frac{5}{4}$

3. $y = -\frac{5}{4}x$

4. $m = -\frac{5}{4}$

5. Create a line with the same slope. Students graphs will vary but should be parallel.

6. Answers will vary. Possible answer: $y = -\frac{5}{4}x + 10$

7. Answers will vary.

8. Answers will vary. Possible answer: $y = -\frac{5}{4}x - 8$

Materials List/Setup

Station 1 graph paper; ruler; red marker; blue marker
Station 2 protractor
Station 3 uncooked spaghetti noodles; graph paper; ruler
Station 4 graph paper; ruler

Expressing Geometric Properties with Equations
Set 1: Parallel Lines, Slopes, and Equations

Instruction

Discussion Guide

To support students in reflecting on the activities and to gather some formative information about student learning, use the following prompts to facilitate a class discussion to "debrief" the station activities.

Prompts/Questions

1. If two lines have the same slope, are they parallel or perpendicular?
2. If two lines have slopes that are opposite reciprocals, are they parallel or perpendicular?
3. How can you prove that two lines are parallel?
4. How can you write equations of lines parallel to a given line?

Think, Pair, Share

Have students jot down their own responses to questions, then discuss with a partner (who was not in their station group), and then discuss as a whole class.

Suggested Appropriate Responses

1. parallel
2. perpendicular
3. Demonstrate that they have the same slope, or that they have corresponding angles when cut by a transversal.
4. Find the slope of the line. Use the point-slope form to find equations of parallel lines.

Possible Misunderstandings/Mistakes

- Not realizing that parallel lines have equal slope
- Not realizing that perpendicular lines have slopes that are opposite reciprocals of each other
- Not recognizing that perpendicular lines create 90° angles which can be used as corresponding angles of parallel lines
- Not recognizing that the equations of parallel lines must have the same slope

NAME: _____

Expressing Geometric Properties with Equations
Set 1: Parallel Lines, Slopes, and Equations

Station 1

At this station, you will find graph paper, a ruler, a red marker, and a blue marker. As a group, construct an *x*- and *y*-axis on your graph paper.

1. On your graph paper, construct a straight line that passes through (–2, –5) and (7, 1). What is the slope of this line? _____

   Show how you found this slope by using the red and blue markers to represent rise and run on your graph.

2. Construct a second line that passes through (0, 4) and (18, 16).

   What is the slope of this line? _____

   Show how you found this slope by using the red and blue markers to represent rise and run on your graph.

3. Construct a third line that passes through (0, 0) and (10, 8).

   What is the slope of this line? _____

   Show how you found this slope by using the red and blue markers to represent rise and run on your graph.

4. Of the three straight lines you created, which two lines do you think are parallel? Explain your answer.

continued

Expressing Geometric Properties with Equations
Set 1: Parallel Lines, Slopes, and Equations

For problems 5–8, determine whether the lines are parallel. Answer *yes* or *no*. Justify your answers.

5.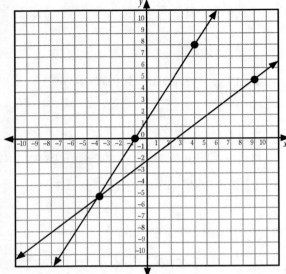

   Are the lines parallel? _____

6.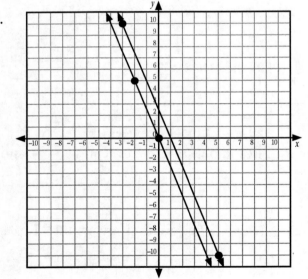

   Are the lines parallel? _____

7. One line has a slope of $\dfrac{2}{3}$ and another line has a slope of $\dfrac{16}{24}$.

   Are the lines parallel? _____

8. One line has a slope of $-\dfrac{1}{2}$ and another line has a slope of $\dfrac{1}{2}$.

   Are the lines parallel? _____

NAME:

Expressing Geometric Properties with Equations
Set 1: Parallel Lines, Slopes, and Equations

Station 2

At this station, you will find a protractor. Work as a group to answer the questions.

Below are two lines cut by a transversal.

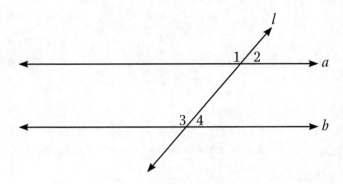

The lines look parallel, but are they?

1. Which pairs of angles are corresponding angles? Explain your answer.

2. Use your protractor to measure angles 1–4. Record your answers below.

   $m\angle 1 =$ _____ $m\angle 2 =$ _____ $m\angle 3 =$ _____ $m\angle 4 =$ _____

3. Which pairs of angles have equal measure? _____

4. If corresponding angles have equal measure, then the lines are parallel. Are these lines parallel? Why or why not?

continued

Expressing Geometric Properties with Equations
Set 1: Parallel Lines, Slopes, and Equations

Let's say you didn't have a protractor, but were given the angle measures shown below.

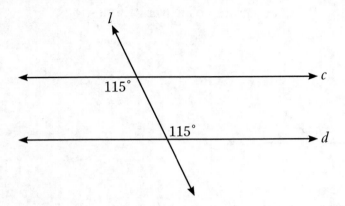

5. How could you use your knowledge of supplementary and corresponding angles to prove that the two lines c and d are parallel?

6. How could you use your knowledge of vertical and corresponding angles to prove that the two lines are parallel?

Expressing Geometric Properties with Equations
Set 1: Parallel Lines, Slopes, and Equations

Station 3

At this station, you will find spaghetti noodles, graph paper, and a ruler. Work as a group to construct the lines and answer the questions.

1. On your graph paper, construct a straight line that passes through (–2, –2) and (5, 10). Place a spaghetti noodle on this line.

2. On the same graph, construct a straight line that passes through (–6, 4) and (1, 16). Place a spaghetti noodle on this line.

By looking at the spaghetti noodles, the two lines may appear to be parallel, but are they?

3. How can you use the points on each line to determine if the two lines are parallel? Show your work and answer *yes* or *no* on the line provided.

    Are the lines parallel? _____

4. On the same graph, construct a straight line that passes through (5, 10) and (17, 3). Place a spaghetti noodle on this line.

    What is the slope of this line? _____

    Is this line perpendicular to the line you created in problem 1? Why or why not?

    Is this line also perpendicular to the line you created in problem 2? Why or why not?

continued

Expressing Geometric Properties with Equations
Set 1: Parallel Lines, Slopes, and Equations

5. Perpendicular lines create four angles that each measure _____.

6. Using your knowledge of corresponding angles and parallel lines, are the lines you created in problems 1 and 2 parallel? Why or why not?

7. Based on your observations in problems 1–6, if two lines are _____ to the same line, then the two lines are _____.

NAME: _____

Expressing Geometric Properties with Equations
Set 1: Parallel Lines, Slopes, and Equations

Station 4

At this station, you will find graph paper and a ruler. Work as a group to construct the graphs and answer the questions.

1. On your graph paper, graph the straight line that passes through points (–4, 5) and (4, –5).

2. What is the slope of this line? _____

3. Use the formula for point-slope form to find an equation for this line. Show your work and answer in the space below.

    Point-slope form: $y - y_1 = m(x - x_1)$

    Equation for the line: _____

4. If you were to construct a line parallel to the line in problem 1, what slope would this line have? Explain your answer.

5. How can you use your graph paper and the slope you found in problem 2 to construct a line parallel to the line in problem 1?

    Draw this line on your graph paper.

    continued

Expressing Geometric Properties with Equations
Set 1: Parallel Lines, Slopes, and Equations

6. Use the formula for point-slope form to find an equation for this line. Show your work and answer in the space below.

    Point-slope form: $y - y_1 = m(x - x_1)$

    Equation for the line: _____

7. Construct another parallel line on your graph paper.

8. Use the formula for point-slope form to find an equation for this line. Show your work and answer in the space below.

    Point-slope form: $y - y_1 = m(x - x_1)$

    Equation for the line: _____

Expressing Geometric Properties with Equations

Set 2: Perpendicular Lines

Instruction

Goal: To provide opportunities for students to develop concepts and skills related to finding the distance between a point and line, determining whether two lines are perpendicular, and writing equations for perpendicular lines

Common Core Standards

Congruence

Experiment with transformations in the plane.

G-CO.1. Know precise definitions of angle, circle, perpendicular line, parallel line, and line segment, based on the undefined notions of point, line, distance along a line, and distance around a circular arc.

Make geometric constructions.

G-CO.12. Make formal geometric constructions with a variety of tools and methods (compass and straightedge, string, reflective devices, paper folding, dynamic geometric software, etc.).

Expressing Geometric Properties with Equations

Use coordinates to prove simple geometric theorems algebraically.

G-GPE.4. Use coordinates to prove simple geometric theorems algebraically.

G-GPE.5. Prove the slope criteria for parallel and perpendicular lines and use them to solve geometric problems (e.g., find the equation of a line parallel or perpendicular to a given line that passes through a given point).

Student Activities Overview and Answer Key

Station 1

Students will be given graph paper and a ruler. Students will construct a line and a point using the graph paper and ruler. They will construct vertical and horizontal lines. Then they will use a perpendicular line to find the distance between the line and the point.

Expressing Geometric Properties with Equations
Set 2: Perpendicular Lines

Instruction

Answers

1. horizontal line

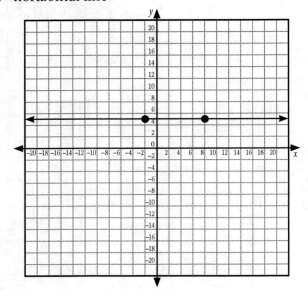

2.

3. Construct a perpendicular line between the horizontal line and the point.
4. 7 units

Expressing Geometric Properties with Equations
Set 2: Perpendicular Lines

Instruction

5. vertical line

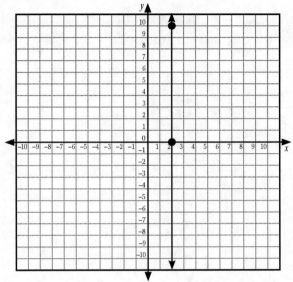

6.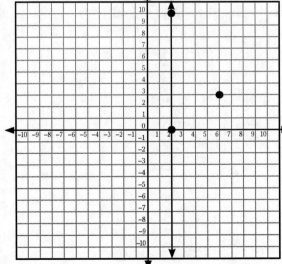

7. Construct a perpendicular line between the vertical line and the point.

8. 4 units

Expressing Geometric Properties with Equations
Set 2: Perpendicular Lines

Instruction

Station 2

Students will be given graph paper, a ruler, and a calculator. Students will use the graph paper and ruler to construct a line and point. They will use the distance formula to find the distance between the point and the line.

Answers

1. $m = -1/2$

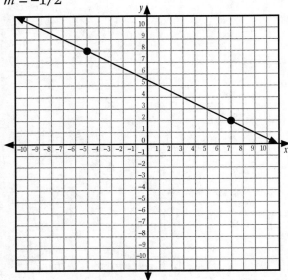

2. $y = \dfrac{-1}{2}x + \dfrac{11}{2}; \; x + 2y = 11$

3.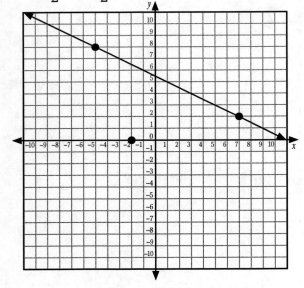

4. $d = \left| \dfrac{1(-2) + 2(0) + (-11)}{\sqrt{1^2 + 2^2}} \right| = \dfrac{13}{\sqrt{5}} = 5.81 \text{ units}$

Expressing Geometric Properties with Equations
Set 2: Perpendicular Lines

Instruction

5. $m = 2$
6. Yes, just over 5 units.
7. It won't be accurate; since the line is not vertical or horizontal, there will be a fraction of units in the distance.

Station 3

Students will be given a protractor. Students will use slope to determine whether or not two lines are perpendicular given the lines on a coordinate plane. Then they will use a protractor to determine whether two lines are perpendicular.

Answers

1. Yes, because the slopes are $m = 1$ and $m = -1$, which are opposite reciprocals.
2. No, because the slopes are $m = 4/5$ and $m = -5/2$, which are not opposite reciprocals.
3. I found the slope of each line. Slopes of perpendicular lines are opposite reciprocals of each other.
4. No, because the angles between the lines are 100° and 80°.
5. Yes, because the angles between the lines are 90°.
6. I measured the angles between the two lines. If all the angles are 90°, then the lines are perpendicular.

Station 4

Students will be given graph paper and a ruler. Students will construct a straight line. They will find the slope of this line and use it to find the slope of a line perpendicular to it. Then they will use the point-slope form of a linear equation to find the equation of the perpendicular line. They will justify their answer by graphing the perpendicular lines on a coordinate plane.

Answers

1. $m = 2$
2. $m = -1/2$; perpendicular lines have slopes that are opposite reciprocals of each other.
3. $y = -\dfrac{1}{2}x + 3$
4. $y = \dfrac{6}{5}x + \dfrac{4}{5}$

Expressing Geometric Properties with Equations
Set 2: Perpendicular Lines

Instruction

5.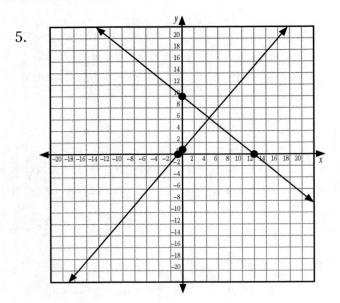

Materials List/Setup

Station 1 graph paper; ruler
Station 2 graph paper; ruler; calculator
Station 3 protractor
Station 4 graph paper; ruler

Expressing Geometric Properties with Equations
Set 2: Perpendicular Lines

Instruction

Discussion Guide

To support students in reflecting on the activities and to gather some formative information about student learning, use the following prompts to facilitate a class discussion to "debrief" the station activities.

Prompts/Questions

1. How can you find the distance between a point and a horizontal or vertical line?
2. How can you find the distance between a point and a line that is not horizontal or vertical?
3. What are two methods you can use to determine if two lines are perpendicular?
4. How can you write an equation for a line that is perpendicular to a given line and passes through a given point?

Think, Pair, Share

Have students jot down their own responses to questions, then discuss with a partner (who was not in their station group), and then discuss as a whole class.

Suggested Appropriate Responses

1. Create a perpendicular line between the point and line. Find the distance of this perpendicular line.
2. Find the equation of the line and write the equation in standard form. Then use the distance formula with the equation and the point to find the distance.
3. Perpendicular lines have opposite reciprocal slopes. Perpendicular lines create four 90° angles.
4. Find the equation of the original line. The slope of the perpendicular line will be the opposite reciprocal of the slope of the given line. Use the point-slope form of a linear equation and the slope of the perpendicular line to find the equation of the perpendicular line.

Possible Misunderstandings/Mistakes

- Not realizing that perpendicular lines have opposite reciprocal slopes
- Not realizing that perpendicular lines create four 90° angles
- Not understanding how to use the point-slope form of a linear equation
- Incorrectly finding the slope of a line as $\frac{run}{rise}$ instead of $\frac{rise}{run}$

Expressing Geometric Properties with Equations
Set 2: Perpendicular Lines

Station 1

At this station, you will find graph paper and a ruler. As a group, construct an *x*- and *y*-axis on the graph paper.

1. Construct a line that passes through (–2, 5) and (8, 5).

   What type of line have you created? _____

2. On the same graph, plot and label the point (3, 12).

3. How can you use perpendicular lines to find the distance between the line you created in problem 1 and the point you plotted in problem 2?

4. What is the distance between the line you created in problem 1 and the point you plotted in problem 2? Explain your answer.

5. On a new graph, construct a line that passes through (2, 10) and (2, 0).

   What type of line have you created? _____

6. On the same graph, plot and label the point (6, 3).

continued

Expressing Geometric Properties with Equations
Set 2: Perpendicular Lines

7. How can you use perpendicular lines to find the distance between the line you created in problem 5 and the point you plotted in problem 6?

8. What is the distance between the line you created in problem 5 and the point you plotted in problem 6? Explain your answer.

Expressing Geometric Properties with Equations
Set 2: Perpendicular Lines

Station 2

At this station, you will find graph paper, a ruler, and a calculator. As a group, create an *x*- and *y*-axis on your graph paper. Work together to answer the questions.

The distance from a point (*m*, *n*) to the line $Ax + By + C = 0$ is given by:

$$d = \left| \frac{Am + Bn + C}{\sqrt{A^2 + B^2}} \right|$$

1. On your graph paper, construct a straight line that passes through (–5, 8) and (7, 2).

    What is the slope of this line? _____

2. What is the equation for this line? _____

    Write this equation in standard form: _____

3. On the same graph, plot and label the point (–2, 0).

4. Use the given distance formula and calculator to find the distance between the line and the point (–2, 0). Show your work and answer in the space below.

    Distance: _____

5. What is the slope of a line perpendicular to the line you graphed in problem 1?

    _____

continued

Expressing Geometric Properties with Equations
Set 2: Perpendicular Lines

6. You can check the reasonableness of your answer to problem 4 by sketching a perpendicular line from the line to the point. *Hint:* Use the slope of the perpendicular line from your point to the line given in problem 1.

   Is the distance of this perpendicular line close to the answer you found in problem 4?
   _____

7. Why can't you just count the units on the perpendicular line between the point and the line?

Expressing Geometric Properties with Equations
Set 2: Perpendicular Lines

Station 3

At this station, you will find a protractor.

For problems 1 and 2, work as a group to determine whether or not the two lines are perpendicular using the coordinate graph. Do NOT use your protractor.

1.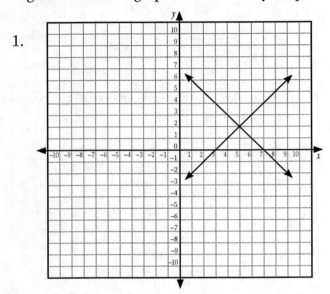

Are the lines perpendicular? Explain.

2.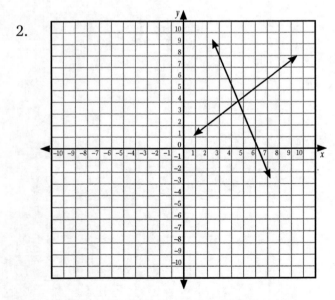

Are the lines perpendicular? Explain.

continued

Expressing Geometric Properties with Equations
Set 2: Perpendicular Lines

3. What strategy did you use to determine whether or not the lines are perpendicular?

For problems 4 and 5, use your protractor to determine whether or not the lines are perpendicular.

4.

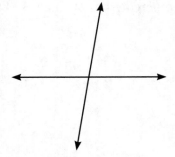

Are the lines perpendicular? Explain.

5.

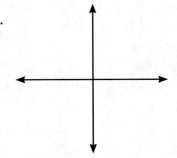

Are the lines perpendicular? Explain.

6. How did you use your protractor to determine whether or not the lines were perpendicular?

Expressing Geometric Properties with Equations
Set 2: Perpendicular Lines

Station 4

At this station, you will find graph paper and a ruler. As a group, create an *x*- and *y*-axis on the graph paper. Work together to answer the questions.

1. On your graph paper, construct a line that passes through points (–4, 0) and (0, 8).

    What is the slope of this line? _____

2. What is the slope of a line that is perpendicular to the line you created in problem 1? Explain your answer.

3. How can you use the point-slope form of an equation, $y - y_1 = m(x - x_1)$, to find the equation of a line perpendicular to the line from problem 1 and passes through the point (–2, 4)? Show your work and answer in the space below.

    Equation: _____

4. Write the equation for the line that is perpendicular to $y = -\frac{5}{6}x + 10$ and passes through (1, 2). Show your work and answer in the space below.

    Equation: _____

5. On your graph paper, graph the two lines in problem 4 to justify that the lines are perpendicular.

Expressing Geometric Properties with Equations

Set 3: Coordinate Proof with Quadrilaterals

Instruction

Goal: To provide opportunities for students to develop concepts and skills related to using coordinate geometry to prove properties of congruent, regular, and similar quadrilaterals

Common Core Standards

Congruence

Experiment with transformations in the plane.

G-CO.1. Know precise definitions of angle, circle, perpendicular line, parallel line, and line segment, based on the undefined notions of point, line, distance along a line, and distance around a circular arc.

Expressing Geometric Properties with Equations

Use coordinates to prove simple geometric theorems algebraically.

G-GPE.4. Use coordinates to prove simple geometric theorems algebraically.

Student Activities Overview and Answer Key

Station 1

Students will be given graph paper and a ruler. Students will construct a trapezoid using coordinate geometry. They will construct trapezoids that are congruent and not congruent to the original trapezoid. Then they will construct a kite using coordinate geometry and determine whether another kite is congruent to the original kite.

Answers

1. trapezoid
2. Answers will vary; corresponding sides are congruent.
3. Answers will vary; corresponding sides are NOT congruent.
4. kite
5. No, because corresponding sides are not congruent; answers will vary.

Expressing Geometric Properties with Equations
Set 3: Coordinate Proof with Quadrilaterals

Instruction

Station 2

Students will be given graph paper and a ruler. Students will construct a square and a rectangle using coordinate geometry. They will prove that squares are regular quadrilaterals. They will show that a square is also a rhombus and a rectangle. They will show that only quadrilaterals that are squares are classified as regular quadrilaterals.

Answers

1. Square, 4 congruent sides, 4 congruent angles; rhombus, 4 congruent sides; rectangle, all squares are rectangles.
2. rectangle
3. Yes, because all sides are congruent and all angles are congruent.
4. No, because not all sides are congruent.
5. Answers will vary.
6. four congruent sides and four congruent angles
7. No, only quadrilaterals that are squares are classified as regular quadrilaterals.

Station 3

Students will be given graph paper and a ruler. Students will construct similar parallelograms, trapezoids, and kites using coordinate geometry. They will explain why different quadrilaterals are similar or not similar. Then they will provide an example of similar quadrilaterals used in the real world.

Answers

1.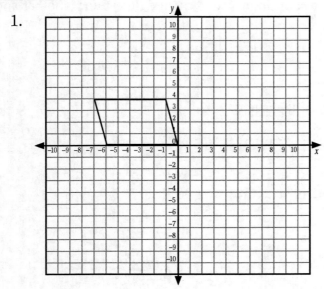

Expressing Geometric Properties with Equations
Set 3: Coordinate Proof with Quadrilaterals

Instruction

2. Answers will vary; answers will vary; they have the same shape, but different sizes.

3. They are similar because they have the same shape, but different sizes.

4. Answers will vary; answers will vary; same shape and different sizes

5. Answers will vary; possible answer: scale drawings

Station 4

Students will be given a real-world application of quadrilaterals and coordinate geometry. Students will identify congruent, similar, and real quadrilaterals. They will identify the name of each quadrilateral.

Answers

1. English and history; science and math; they have the same size and shape.

2. Café and library; dormitory and pool; they have the same shape, but different size.

3. Yes; the café and library because all sides are congruent and all angles are congruent

4. square and rhombus and rectangle; square and rhombus and rectangle; parallelogram; trapezoid; trapezoid; kite; kite; parallelogram

Materials List/Setup

Station 1   graph paper; ruler
Station 2   graph paper; ruler
Station 3   graph paper; ruler
Station 4   none

Expressing Geometric Properties with Equations
Set 3: Coordinate Proof with Quadrilaterals

Instruction

Discussion Guide

To support students in reflecting on the activities and to gather some formative information about student learning, use the following prompts to facilitate a class discussion to "debrief" the station activities.

Prompts/Questions

1. How can you use coordinate geometry to prove two quadrilaterals are congruent?
2. How can you use coordinate geometry to prove two quadrilaterals are similar?
3. How can you use coordinate geometry to prove a quadrilateral is "regular"?
4. What are some real-world applications of quadrilaterals and coordinate geometry?

Think, Pair, Share

Have students jot down their own responses to questions, then discuss with a partner (who was not in their station group), and then discuss as a whole class.

Suggested Appropriate Responses

1. Use the units in the coordinate graph to show that the quadrilaterals have congruent corresponding sides and congruent corresponding angles.
2. Use the units in the coordinate graph to show that the quadrilaterals have the same shape, but are different sizes.
3. Use the units to show that all sides of the quadrilateral are congruent and all angles of the quadrilateral are congruent.
4. Possible answers: scale drawings, art and design, stained glass making, architecture

Possible Misunderstandings/Mistakes

- Not recognizing that the units in a graph can determine the side lengths of quadrilaterals
- Not recognizing that the units in a graph can determine whether two quadrilaterals are congruent or similar
- Not realizing that a "regular" quadrilateral must have four equal sides and four equal angles

NAME:

Expressing Geometric Properties with Equations
Set 3: Coordinate Proof with Quadrilaterals

Station 1

At this station, you will find graph paper and a ruler. Work as a group to construct the quadrilaterals and answer the questions.

1. On your graph paper, construct a quadrilateral that has vertices (1, 1), (2, 4), (8, 4), and (9, 1).

   What type of quadrilateral did you create? _____

2. On the same graph, construct a quadrilateral that is congruent to the quadrilateral in problem 1.

   What are the vertices for this new quadrilateral? _____

   Explain why the quadrilaterals in problems 1 and 2 are congruent.

3. On the same graph, construct a quadrilateral that is NOT congruent to the quadrilateral in problem 1.

   What are the vertices of this new quadrilateral? _____

   Explain why the quadrilaterals in problems 1 and 3 are NOT congruent.

4. Graph a quadrilateral with vertices (0, 0), (2, 5), (–2, 5), and (0, 7).

   What type of quadrilateral did you create? _____

continued

Expressing Geometric Properties with Equations
Set 3: Coordinate Proof with Quadrilaterals

5. Jacob claims that a quadrilateral with vertices (–2, 2), (0, 6), (–2, 8), and (–4, 6) is congruent to the quadrilateral in problem 4. Is Jacob correct? Why or why not? Graph the quadrilaterals on your graph paper to justify your answer.

If Jacob is incorrect, what vertices would make this quadrilateral congruent to the quadrilateral in problem 4?

Expressing Geometric Properties with Equations
Set 3: Coordinate Proof with Quadrilaterals

Station 2

At this station, you will find graph paper and a ruler. Work as a group to construct the quadrilaterals and answer the questions.

1. On your graph paper, graph a quadrilateral that has vertices (2, 4), (8, 4), (2, 10), and (8, 10).

   What are three names for this quadrilateral? Justify each name using properties of that type of quadrilateral.

2. On your graph paper, graph a quadrilateral that has vertices (–1, 0), (–4, 0), (–1, 7), and (–4, 7).

   What is the name of this quadrilateral? _____

3. Is the quadrilateral in problem 1 a regular quadrilateral? Why or why not?

4. Is the quadrilateral in problem 2, a regular quadrilateral? Why or why not?

5. What strategy did you use for problems 1–4?

continued

Expressing Geometric Properties with Equations
Set 3: Coordinate Proof with Quadrilaterals

6. What is the definition of a regular quadrilateral?

7. Are all rhombi and rectangles regular quadrilaterals? Why or why not?

Expressing Geometric Properties with Equations
Set 3: Coordinate Proof with Quadrilaterals

Station 3

At this station, you will find graph paper and a ruler. Work together to construct the quadrilaterals and answer the questions.

1. On your graph paper, construct a parallelogram with vertices (0, 0), (–6, 0), (–7, 4), and (–1, 4).

2. On the same graph, construct a similar parallelogram.

    What are the vertices of the parallelogram? _____

    Why is this new parallelogram similar to the parallelogram in problem 1?

3. On a new graph, construct a trapezoid with vertices (1, 1), (4, 1), (2, 4), and (3, 4). Construct a second trapezoid with vertices (0, 0), (6, 0), (2, –6), and (4, –6).

    Are the two trapezoids congruent or similar? Explain your answer.

4. On a new graph, construct a kite. Construct a second kite that is similar, but not congruent to the first kite.

    What are the vertices of the first kite? _____

    What are the vertices of the second kite? _____

    Why are the kites similar?

5. When would you use similar quadrilaterals in the real world?

Expressing Geometric Properties with Equations
Set 3: Coordinate Proof with Quadrilaterals

Station 4

Below is the map of a college campus. Use what you know about quadrilaterals and coordinate geometry to answer the questions about the map.

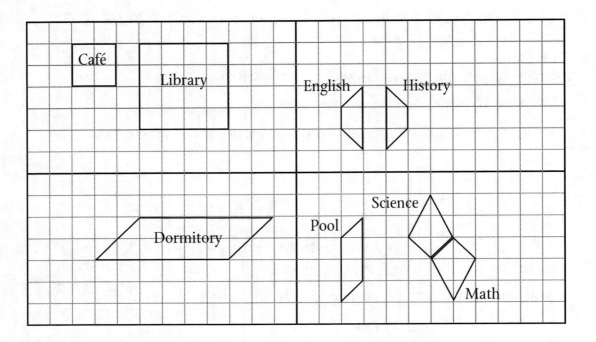

1. Which buildings are congruent? Justify your answer.

2. Which buildings are similar? Justify your answer.

3. Are there any regular quadrilaterals in the map? Explain your answer.

continued

Expressing Geometric Properties with Equations
Set 3: Coordinate Proof with Quadrilaterals

4. What type of quadrilateral is each building? (List all possible names for each building.)

   Café: _____

   Library: _____

   Dormitory: _____

   English building: _____

   History building: _____

   Science building: _____

   Math building: _____

   Pool: _____